Lecture Notes in Mathematics

1692

Editors:
A. Dold, Heidelberg
F. Takens, Groningen
B. Teissier, Paris

Springer
Berlin
Heidelberg
New York
Barcelona
Budapest
Hong Kong
London
Milan
Paris
Singapore
Tokyo

Timothy M. W. Eyre

Quantum Stochastic Calculus and Representations of Lie Superalgebras

Springer

Author

Timothy M. W. Eyre
Data Connection Ltd
100 Church Street
Enfield, EN2 6BQ, UK
e-mail: tme@datcon.co.uk

Cataloging-in-Publication Data applied for

Die Deutsche Bibliothek - CIP-Einheitsaufnahme

Eyre, Timothy M. W.:
Quantum stochastic calculus and representations of Lie
superalgebras / Timothy M. W. Eyre. - Berlin ; Heidelberg ; New
York ; Barcelona ; Budapest ; Hong Kong ; London ; Milan ; Paris ;
Santa Clara ; Singapore ; Tokyo : Springer, 1998
  (Lecture notes in mathematics ; 1692)
  ISBN 3-540-64897-6

Mathematics Subject Classification (1991): 81S25

ISSN 0075-8434
ISBN 3-540-64897-6 Springer-Verlag Berlin Heidelberg New York

Typesetting: Camera-ready TeX output by the author
SPIN: 10649929   41/3143-543210 - Printed on acid-free paper

Preface

These notes are intended to give a self-contained exposition of $\mathbf{Z}_2$-graded quantum stochastic calculus and the resulting representations of Lie superalgebras. No pre-knowledge of quantum stochastic calculus is assumed; Chapter 2 presents the Hudson-Parthasarathy formulation of the theory in a straightforward manner.

I should like to thank Professor Robin Hudson for his collaboration during the years 1994-1997 which I spent at The University of Nottingham where this research was carried out. I should also like to thank EPSRC for funding this research.

Special thanks must go to the Indian Statistical Institute New Delhi Centre, India (especially K. R. Parthasarathy), the Wydzial Matematyki, University of Lódź, Poland (especially Stanislaw Goldstein) and the Institute Fourier, Grenoble, France (especially Stephane Attal) for their kind hospitality. These academic visits were all of great value to the research.

Table of Contents

1. Introduction

1.1 Discussion

A theory of quantum stochastic calculus was devised in [HP1]. This theory was based on the canonical commutation relations of quantum theory and hence is a *Bosonic* quantum stochastic calculus. A Fermionic version of quantum stochastic calculus, based on the canonical anticommutation relations, was given in [AH] with [BSW] giving an earlier version. The Bosonic and Fermionic versions of quantum stochastic calculus were unified for the one-dimensional case in [HP2] by means of the formula

$$dB = (-1)^A dA. \tag{1.1}$$

The full Ito algebra of one-dimensional Fermionic quantum stochastic calculus consists of the complex span of the quantum stochastic differentials dA, dT, dB and $dB^\dagger$. The differential dA corresponds to the one-dimensional gauge process, dT to the time process, dB to the Fermionic annihilation process and $dB^\dagger$ to the Fermionic creation process. Of these differentials, dA and dT commute about integrands whereas dB and $dB^\dagger$ commute or anticommute about an integrand X depending on whether $(-1)^A X (-1)^A = X$ or $-X$ respectively.

It therefore makes sense to $\mathbf{Z}_2$-grade the Ito algebra of Fermionic quantum stochastic calculus by means of the assignments dB, $dB^\dagger$ odd, dA, dT even. The Fermionic quantum Ito formula given in [AH] shows that, equipped with this grading, the Fermionic Ito algebra satisfies the conditions for a $\mathbf{Z}_2$-graded algebra. The general theory of $\mathbf{Z}_2$-graded algebras is described in [C].

Here we describe an extension of (1.1) which gives rise to a $\mathbf{Z}_2$-graded theory of multidimensional quantum stochastic calculus. In this theory we have a mixture of multidimensional Bosonic and Fermionic creation and annihilation processes along with $\mathbf{Z}_2$-graded multidimensional gauge processes. The commutation and anticommutation properties of these processes are shown to provide a time-indexed family of representations of a broad class of Lie superalgebras. This result is by no means one that is to be expected or even conjectured. The relation

$$\{\Xi_A, \Xi_B\} = \Xi_{\{A,B\}}$$

summarises this representation result where $\{\cdot\,,\cdot\}$ denotes the supercommutator bracket and $\varXi_A$, $\varXi_B$ are $\mathbf{Z}_2$-graded quantum stochastic integrator processes. This work has been published in our paper [EH].

A particular case of this result is that all Lie superalgebras of the form $gl(N, r)$ may be represented in $\mathbf{Z}_2$-graded quantum stochastic calculus by restricting attention to the $\mathbf{Z}_2$-graded multidimensional pure gauge processes. It is natural to suppose that representation results such as this will lead to productive links between quantum stochastic calculus and the representation theory of graded structures.

The Lie superalgebra representation result is used in these notes to provide a formula for the chaotic decomposition of elements of the universal enveloping superalgebra of the Lie superalgebra associated with $\mathbf{Z}_2$-graded quantum stochastic calculus. This is done by exploiting the co-(super)algebra structure of the universal enveloping superalgebra.

A quantum group is a non-commutative non-co-commutative Hopf algebra. Such an algebra is normally obtained by twisting the co-product of the infinite dimensional universal enveloping algebra of a semi-simple Lie algebra by a group-like element. In the case where the Lie algebra is sl_2 this leads to a non-trivial deformation of the universal enveloping algebra. Such theories also exist for the Lie superalgebra case. The effectiveness with which the co-algebra structure of a universal enveloping superalgebra is used in Chapter 8 suggests that $\mathbf{Z}_2$-graded quantum stochastic calculus may provide useful results in the theory of quantum groups. This is an area of research which has seen much recent activity and some details of the subject can be found in [CP].

Co-algebra structures are used in the pioneering paper [Sch1]. Here it is seen that quantum independent increment processes are related to quantum stochastic differential equations. Further work in this area is given in [Sch2].

1.2 Overview

Chapter 2 consists of a review of the Hudson-Parthasarathy formulation of quantum stochastic calculus. This is crucial as the results that follow depend heavily on the results of Hudson and Parthasarathy. The standard reference work for quantum stochastic calculus is [P1]. A less demanding treatment of this area can be found in R. L. Hudson's unpublished lecture notes (1994) and a brief exposition intended for a general audience is given in [P3]. The first section of Chapter 2 constructs Fock space by means of the kernel construction rather than by the perhaps more widely-known symmetrised tensor product method. Section 2.2 defines the integrator processes of quantum stochastic processes in Fock space. These processes are given for the N-dimensional case and are known as the N-dimensional creation, annihilation and gauge processes. From these processes, simple quantum stochastic integrals are constructed in Section 2.3. Here two important results of [HP1] are stated known as the first and second fundamental formulas. The first fundamental formula provides the matrix elements of a quantum stochastic integral. The second fundamental formula provides a notion of multiplication of quantum stochastic integrals in spite of their unboundedness. In Section 2.4 these simple integrals are extended by means of the fundamental estimate. It is stated that the first and second fundamental formulas continue to hold for extended integrals. In Section 2.5 it is shown that quantum stochastic integrals are themselves processes which are susceptible to quantum stochastic integration and so iterated quantum stochastic integrals may be defined. The chapter ends with a section devoted to a brief review of the Boson–Fermion unification result of [HP2].

Chapter 3 reviews the theory of $\mathbf{Z}_2$-graded structures and provides the algebraic machinery that will be required later. Section 3.1 describes the notion of a $\mathbf{Z}_2$-graded vector space and introduces for the first time the important graded tensor product. In Section 3.2 the notion of a $\mathbf{Z}_2$-grading is extended to algebras and some important technical definitions are given. A $\mathbf{Z}_2$-graded algebra is commonly known as a *superalgebra*. The definition of a Lie superalgebra is given in Section 3.3. To any Lie superalgebra there corresponds an important entity known as the *universal enveloping superalgebra*. This is the $\mathbf{Z}_2$-graded analogue of the universal enveloping algebra corresponding to a Lie algebra. The definition of the universal enveloping superalgebra associated with a Lie superalgebra along with some useful related results is given in Section 3.4. Details of the theory of graded structures can be found in [B,C]. The notion of a Lie superalgebra was introduced in [K] and another work on the subject is [S]. Works on Lie algebras include [Hum,J] and [D] is devoted to enveloping algebras.

Chapter 4 contains the central result of this book. The material here is derived from our paper [EH]. Section 4.1 discusses the Lie algebra representations that are obtained from ungraded quantum stochastic calculus. In Section 4.2 a second review of Boson–Fermion equivalence [HP2] is given, this time from the point of view of the theory that is to follow. A partial generalisation to N dimensions of this equivalence is given in Section 4.3. The full $\mathbf{Z}_2$-grading of multidimensional quantum stochastic calculus begins in Section 4.4 where the indexing algebra of the quantum stochastic differentials is $\mathbf{Z}_2$-graded. The differentials and processes of $\mathbf{Z}_2$-graded quantum stochastic calculus are defined in Section 4.5. Some technical lemmas are stated and proved in Section 4.6. In Section 4.7 the theorem that underlies much of $\mathbf{Z}_2$-graded quantum stochastic calculus is stated and proved. This theorem states that the processes of $\mathbf{Z}_2$-graded quantum stochastic calculus provide a time-indexed family of representations of a broad class of Lie superalgebras.

In Chapter 5 a notion of multiplication of iterated ungraded quantum stochastic integrals is given. This theory has been published in [CEH].

The $\mathbf{Z}_2$-graded analogue of the theory given in Chapter 5 is begun in Chapter 6. Section 6.1 gives some preliminary definitions. In particular, it defines the Ito superalgebra $\mathcal{I}$. This Ito superalgebra is taken to consist of quantum stochastic differentials in later chapters but in Chapter 6 the theory is entirely abstract. The Ito superalgebra $\mathcal{I}$ is put to use in Chapter 7 and forms the central object of Chapter 8. In Section 6.2 a technical product is defined that will lead to the definition of the important $\star$ product which is defined on the tensor space of $\mathcal{I}$ in Section 6.4. Section 6.3 is devoted to the description of a useful computational device which simplifies the calculations of Section 6.2. This computational device can also be used to provide heuristic insights into various aspects of the $\mathbf{Z}_2$-graded algebraic theory under consideration. Unfortunately, the device can not be considered to be mathematically rigorous and so must be confined to its own section. See, however, [HPu]. Section 6.5 defines supersymmetric tensors and shows that the space of such tensors is closed under the $\star$ product. While this theory is scarcely used in later chapters, the results and proofs are of intrinsic interest and may prove useful in future research.

Chapter 7 provides some useful results concerning $\mathbf{Z}_2$-graded quantum stochastic calculus. Section 7.1 derives a version of the first fundamental formula for $\mathbf{Z}_2$-graded quantum stochastic calculus. A similar task is performed for the second fundamental formula in Section 7.2. This version of the second fundamental formula is discussed in Section 7.3 where a useful differential form of the relation is derived. Section 7.4 shows how the adjoint of a $\mathbf{Z}_2$-graded quantum stochastic integral should be taken. In particular, a proposition giving an explicit formula for the adjoint of an iterated quantum stochastic integral is stated and proved. Chapter 7 is connected with Chapter 6 in Section 7.5 where the Ito superalgebra $\mathcal{I}$ of Chapter 6 is fixed as a superalgebra of $\mathbf{Z}_2$-graded quantum stochastic differentials. A map I called the *integrator map* is defined on the tensor space of $\mathcal{I}$ and enables algebraic structures associated with $\mathcal{I}$ to be exploited in the manipulation of quantum stochastic integrals. It is by means of I and the product $\star$ that a notion of multiplying iterated $\mathbf{Z}_2$-graded quantum stochastic integrals is provided in Section 7.6. The important technical result that the integrator map I is injective is established in Section 7.7. This leads to a rigorous algebraic multiplication of iterated $\mathbf{Z}_2$-graded quantum stochastic integrals being defined in Section 7.8.

In Chapter 8 we move from products of iterated integrals to polynomials of integrator processes. The chaotic decomposition of such a polynomial is derived by establishing a rigorous analogue of the heuristic classical relation

$$df(\Lambda) = f(\Lambda + d\Lambda) - f(d\Lambda)$$

where f is an arbitrary polynomial function. This is done in Section 8.4. Sections 8.1 to 8.3 are mainly technical in nature. The work in Chapter 8 is largely a $\mathbf{Z}_2$-graded analogue of that done for the ungraded case in

[HPu]. Work leading up to the theory described in [HPu] may be found in [HP3,HP4,HP5,H2]. Chapters 6, 7 and 8 are all derived from [E].

While the $\mathbf{Z}_2$-graded case is of obvious physical relevance, it is natural to consider the possibility of grading quantum stochastic calculus over other abelian groups. Sections 9.2 to 9.7 explore this idea by developing a theory of $\mathbf{Z}_n$-graded quantum stochastic calculus. Section 9.8 shows that $\mathbf{Z}_n$-graded quantum stochastic calculus will only provide commutation relations in the elegant form of Chapter 4 in the case $n = 2$. To counter this negative result, a grading of quantum stochastic calculus by the abelian group $\mathbf{Z}_n \times \mathbf{Z}_n$ is outlined in Section 9.9 and it is shown that such a grading does in fact provide attractive commutation relations. Section 9.10 shows how quantum stochastic calculus with an infinite number of degrees of freedom may be $\mathbf{Z}_n$-graded. Of course, it is the case $n = 2$ that is of primary interest here. Section 9.1 gives an overview of the whole of Chapter 9. A different kind of extension of the theory of Lie superalgebra representations in $\mathbf{Z}_2$-graded quantum stochastic calculus can be found in [P2].

1.3 Notational Conventions

We denote the natural numbers by $\mathbf{N}$ and take this set to include 0. We denote the real numbers by $\mathbf{R}$ and the set $\{t \in \mathbf{R} : t \geq 0\}$ by $\mathbf{R}_+$. We denote the complex numbers by $\mathbf{C}$.

For arbitrary N in $\mathbf{N}$ with $N \geq 1$ we denote the space of square integrable functions from $\mathbf{R}_+$ to $\mathbf{C}^N$ by $L^2(\mathbf{R}_+; \mathbf{C}^N)$. For an element f of $L^2(\mathbf{R}_+; \mathbf{C}^N)$ and an integer i with $1 \leq i \leq N$ we denote the ith component of f by f^i so that $f = (f^1, \ldots f^N)$. We denote the conjugate $\bar{f}^i$ of f^i by f_i.

Inner products are taken to be linear on the right and conjugate linear on the left.

We will take advantage of the Einstein repeated suffix summation convention. We also adhere to the related convention that Greek indices such as α, β, γ, ... run from 0 whereas Roman indices i, j, k, ... run from 1.

We denote by $\mathbf{Z}_2$ the abelian group consisting of the set $\{0, 1\}$ equipped with modulo 2 addition. It is sometimes illuminating to consider $\mathbf{Z}_2$ as one of its isomorphs such as $\{1, -1\}$ under standard multiplication. We denote the abelian group consisting of the set $\{0, \ldots, n - 1\}$ equipped with the modulo n addition by $\mathbf{Z}_n$.

Matrices will normally be written as capital letters and sometimes in the form $((a))$ where this is appropriate.

If A, B and C are sets and f and g are functions with $f \colon A \to B$ and $g \colon B \to C$ then we denote the composition of f with g by $g \circ f$. For a set A containing B as a subset we denote the indicator function of B by χ_B where

$$\chi_B \colon A \to \mathbf{C}, \qquad \chi_B(a) = \begin{cases} 1 & \text{if } a \in B; \\ 0 & \text{if } a \notin B. \end{cases}$$

All vector spaces given will be over the complex numbers. By 'algebra' we mean nothing more than a complex vector space equipped with a bilinear closed multiplication. All additional properties possessed by an algebra such as associativity, unitality, gradedness or being a Lie algebra will be stated explicitly.

We denote the Hilbert space tensor product by $\otimes$. Where the algebraic tensor product is required it is denoted $\underline{\otimes}$. The graded tensor product which will be introduced will also be denoted $\otimes$.

We denote the group of permutations of a set of cardinality m by S_m.

The expression 'this work' always refers to the entire book. Each chapter is divided into numbered sections. All definitions, lemmas, propositions, theorems and corollaries are numbered together and consecutively within each chapter. The end of a proof is indicated by the symbol ∎.

2. Quantum Stochastic Calculus

2.1 The Kernel Construction and Fock Space

In order to describe the kernel construction we must define the notion of a non-negative definite kernel.

Definition 2.1 If X is a set and $K : X \times X \to \mathbf{C}$ is some function then K is a *non-negative definite kernel* over X if, for each integer n with $n \geq 1$, arbitrary $x_1, \ldots, x_n \in X$ and arbitrary $\alpha_1, \ldots, \alpha_n \in \mathbf{C}$ we have

$$\sum_{j,k=1}^{n} \bar{\alpha}_j \alpha_k K(x_j, x_k) \geq 0.$$

We now give the kernel construction in the form of a theorem.

Theorem 2.2 Let K be a non-negative definite kernel over X. Then there exists a pair $(\mathcal{H}, k)$ comprising a Hilbert space $\mathcal{H}$ and a map $k \colon X \to \mathcal{H}$ such that

    (a) $\{k(x) : x \in X\}$ is total in $\mathcal{H}$;

    (b) $\langle k(x), k(y) \rangle_{\mathcal{H}} = K(x, y)$ for arbitrary $x, y \in X$.

If $(\mathcal{H}', k')$ is a second such pair then there exists a unique Hilbert space isomorphism $U \colon \mathcal{H} \to \mathcal{H}'$ such that $U \circ k = k'$.

The pair $(\mathcal{H}, k)$ is called the *Gel'fand pair* associated with K.

    Let h be a Hilbert space with inner product $\langle ., . \rangle_h$. It is true that $\langle ., . \rangle_h$ is a non-negative definite kernel over h and it can be shown from this that $\exp(\langle ., . \rangle_h)$ is also a non-negative definite kernel over h. We denote the Gel'fand pair associated with the non-negative definite kernel $\exp(\langle ., . \rangle_h)$ by $(\Gamma(h), e(\cdot))$ and call $\Gamma(h)$ the *Fock space* over h. For f in h we call $e(f)$ the *exponential vector* corresponding to f. Thus the Fock space over h is a Hilbert space $\Gamma(h)$ equipped with a total family of vectors $(e(f), f \in h)$ satisfying

$$\langle e(f), e(g) \rangle_{\Gamma(h)} = \exp(\langle f, g \rangle_h)$$

for arbitrary $f, g \in h$.

Theorem 2.3 Let $h = h_1 \oplus \cdots \oplus h_n$ be a Hilbert space direct sum. Then the Fock space $\Gamma(h)$ can be realised in terms of $\Gamma(h_1), \ldots, \Gamma(h_n)$ as

$$\Gamma(h) = \Gamma(h_1) \otimes \cdots \otimes \Gamma(h_n)$$

with

$$e((f^1, \ldots, f^n)) = e(f^1) \otimes \cdots \otimes e(f^n)$$

where each f^j is an element of h_j for $j = 1, \ldots, n$.

We now give three useful properties of Fock space in the form of a proposition.

Proposition 2.4 The following three statements are true:

(i)   The map $f \mapsto e(f)$ from h to $\Gamma(h)$ is continuous and differentiable.
(ii) If $\mathcal{M}$ is a dense subset of h then $\{e(f) : f \in \mathcal{M}\}$ is total in $\Gamma(h)$.
(iii) The exponential vectors are linearly independent in $\Gamma(h)$.

Part (ii) of this proposition follows as a corollary of part (i).

The Fock spaces that are of interest for the purpose of developing quantum stochastic calculus are those of the form $\Gamma(L^2(\mathbf{R}_+; \mathbf{C}^N))$ where N is a non-zero natural number. We define $\mathbf{L}^2(\mathbf{R}_+; \mathbf{C}^N)$ to be the subspace of $L^2(\mathbf{R}_+; \mathbf{C}^N)$ consisting of those elements of $L^2(\mathbf{R}_+; \mathbf{C}^N)$ that are locally bounded. Similarly, if A is a subset of $\mathbf{R}_+$ then we define $\mathbf{L}^2(A; \mathbf{C}^N)$ to be the space of all locally bounded functions in $L^2(A; \mathbf{C}^N)$. It is certainly true that $\mathbf{L}^2(\mathbf{R}_+; \mathbf{C}^N)$ is a dense subset of $L^2(\mathbf{R}_+; \mathbf{C}^N)$ so, by part (ii) of proposition 2.4, we have that the set $\{e(f) : f \in \mathbf{L}^2(\mathbf{R}_+; \mathbf{C}^N)\}$ is total in $\Gamma(L^2(\mathbf{R}_+; \mathbf{C}^N))$.

We denote the space of all finite complex linear combinations of the elements of $\{e(f) : f \in \mathbf{L}^2(\mathbf{R}_+; \mathbf{C}^N)\}$ by $\mathcal{E}$. We refer to $\mathcal{E}$ as the *exponential domain*. Thus we have that

$$\mathcal{E} = \{\lambda_1(f^1) + \cdots + \lambda_n e(f^n) :$$
$$n \in \mathbf{N}, \ \lambda_1, \ldots, \lambda_n \in \mathbf{C}, \ f^1, \ldots, f^n \in \mathbf{L}^2(\mathbf{R}_+; \mathbf{C}^N)\}.$$

Similarly, if A is some subset of $\mathbf{R}_+$ then we denote the set of all finite linear combinations of elements of the set $\{e(f) : f \in \mathbf{L}^2(A; \mathbf{C}^N)\}$ by $\mathcal{E}(A)$.

Since the exponential vectors $\{e(f) : f \in \mathbf{L}^2(\mathbf{R}_+; \mathbf{C}^N)\}$ are linearly independent, they form a basis of $\mathcal{E}$. An operator with domain $\mathcal{E}$ can be defined by specifying an arbitrary action on each exponential vector. We now define a particular variety of operator in $\Gamma(L^2(\mathbf{R}_+; \mathbf{C}^N))$.

Definition 2.5 An *exponential operator* is an operator such that its domain includes the exponential domain $\mathcal{E}$ and such that the domain of its adjoint also includes the exponential domain.

All operators to be considered in this work will be exponential operators. Definition 2.5 may be extended to operators in $\Gamma(L^2(A; \mathbf{C}^N))$ where $A \subset \mathbf{R}_+$ by requiring that the operator and the adjoint of the operator have domains that include $\mathcal{E}(A)$. If X is some exponential operator then we denote the restriction of the adjoint X^* of X to $\mathcal{E}$ by $X^\dagger$ and, for readability, speak of this as the 'adjoint' of X.

The space $L^2(\mathbf{R}_+; \mathbf{C}^N)$ may be decomposed into 'past' and 'future' spaces at an arbitrary time t in $\mathbf{R}_+$ as follows:

$$L^2(\mathbf{R}_+; \mathbf{C}^N) = L^2([0, t]; \mathbf{C}^N) \oplus L^2((t, \infty); \mathbf{C}^N).$$

By theorem 2.3 we have that

$$\Gamma(L^2(\mathbf{R}_+; \mathbf{C}^N)) = \Gamma(L^2([0, t]; \mathbf{C}^N)) \otimes \Gamma(L^2((t, \infty); \mathbf{C}^N)).$$

Thus each exponential vector $e(f)$ with $f \in L^2(\mathbf{R}_+; \mathbf{C}^N)$ factorises as

$$e(f) = e(f_t) \otimes e(f^t)$$

where $f_t = f|_{[0,t]}$ and $f^t = f|_{(t,\infty)}$. Consequently, the exponential domain factorises into the algebraic tensor product

$$\mathcal{E}([0, t]) \underline{\otimes} \mathcal{E}((t, \infty)). \tag{2.1}$$

The product (2.1) can be thought of as the set of all product vectors $\psi \underline{\otimes} \chi$ with $\psi \in \mathcal{E}([0, t])$ and $\chi \in \mathcal{E}((t, \infty))$.

2.2 The Processes of Quantum Stochastic Calculus

A *process* is a family of exponential operators indexed by time. A process $F = (F(t))_{t \in \mathbf{R}_+}$ is said to be *adapted* if each $F(t)$ as an operator on $\mathcal{E} = \mathcal{E}([0, t]) \underline{\otimes} \mathcal{E}((t, \infty))$ is of the form

$$F(t) = F_t \underline{\otimes} 1^t$$

where F_t is some exponential operator in $\mathcal{E}([0, t])$ and 1^t is the identity operator in $\mathcal{E}((t, \infty))$. It is clear that the *adjoint process* $F^\dagger = (F^\dagger(t))_{t \in \mathbf{R}_+}$ of an adapted process F is also an adapted process.

We now wish to define the family of adapted processes around which quantum stochastic calculus is built. First it is necessary to construct the space $M_0(N)$ which is used to index these processes.

In this section, and indeed throughout this work, N continues to denote the index applied to $\mathbf{C}$ in the Fock space $\Gamma(L^2(\mathbf{R}_+; \mathbf{C}^N))$. The space $M_0(N)$ is defined to be the space consisting of all complex $(N+1) \times (N+1)$ matrices indexed from 0 to N. The space $M_0(N)$ is equipped with a multiplication distinct from standard matrix multiplication which will now be described.

Denote by $((\hat{\delta}))$ the matrix that is zero everywhere except for the 1st to N^{th} diagonal entries which take the value 1. Thus we have

$$((\hat{\delta})) = \begin{pmatrix} 0 & 0 & \cdots & 0 \\ 0 & 1 & \cdots & 0 \\ \vdots & \vdots & \ddots & \vdots \\ 0 & 0 & \cdots & 1 \end{pmatrix}.$$

We define the multiplication . in $M_0(N)$ by

$$A.B = A((\hat{\delta}))B \tag{2.2}$$

where A and B are arbitrary elements of $M_0(N)$ and the multiplication in use on the right hand side of (2.2) is standard matrix multiplication.

For arbitrary integers α, β with $0 \le \alpha, \beta \le N$ let E_β^α denote the $(\alpha, \beta)^{th}$ basis element of $M_0(N)$. By this we mean that E_β^α is the $(N+1) \times (N+1)$ matrix with 1 in the entry that occupies the αth column at the βth row and zero in all other entries. From this definition we have for arbitrary integers $\alpha, \beta, \gamma, \delta$ with $0 \le \alpha, \beta, \gamma, \delta \le N$ that

$$E_\beta^\alpha.E_\delta^\gamma = \hat{\delta}_\delta^\alpha E_\beta^\gamma.$$

Here $\hat{\delta}$ denotes the *Evans delta* which is defined for arbitrary integers α, β with $0 \le \alpha, \beta \le N$ as follows:

$$\hat{\delta}_\beta^\alpha = \begin{cases} 1 & \text{if } \alpha = \beta \ne 0; \\ 0 & \text{otherwise.} \end{cases} \tag{2.3}$$

Let e_i denote the ith canonical basis element of $\mathbf{C}^N$, that is to say, e_i denotes the element $(0, \ldots, 0, 1, 0, \ldots, 0)$ of $\mathbf{C}^N$ where the lone '1' occurs in the ith position.

We may now define the family of adapted processes $(\Lambda_A)_{A \in M_0(N)}$ that form the basis of quantum stochastic calculus. We begin by defining for arbitrary integers α, β with $0 \le \alpha, \beta \le N$ the processes $\Lambda_{E_\beta^\alpha}$ which are by convention denoted as Λ_β^α, see [Ev]. It is sufficient to give the action of each Λ_β^α at an arbitrary time t in $\mathbf{R}_+$ on an arbitrary exponential vector $e(f)$ with $f \in L^2(\mathbf{R}_+; \mathbf{C}^N)$. There are four different cases which must be treated separately.

If i, j are such that $1 \le i, j \le N$ then we define Λ_j^i by prescribing that, at an arbitrary time $t \ge 0$, $\Lambda_j^i(t)$ acts on an arbitrary exponential vector $e(f)$ with $f \in L^2(\mathbf{R}_+; \mathbf{C}^N)$ as follows:

$$\Lambda_j^i(t)e(f) = \frac{d}{dz}e(e^{\chi_{[0,t]}z|e_j\rangle\langle e_i|}f)|_{z=0}. \tag{2.4}$$

In (2.4) we denote by $|e_j\rangle\langle e_i|$ the Dirac dyad acting in $L^2(\mathbf{R}_+; \mathbf{C}^N)$. This object maps a function f in $L^2(\mathbf{R}_+; \mathbf{C}^N)$ to the $L^2(\mathbf{R}_+; \mathbf{C}^N)$ function that has zero for all components except the jth component which is f^i, the ith component of f. For arbitrary j with $1 \leq j \leq N$ and arbitrary $t \geq 0$ we define $\Lambda_j^0(t)$ by

$$\Lambda_j^0(t)e(f) = \frac{d}{dz}e(f + \chi_{[0,t]}ze_j)|_{z=0}. \qquad (2.5)$$

For arbitrary i with $1 \leq i \leq N$ we define $\Lambda_0^i(t)$ by

$$\Lambda_0^i(t)e(f) = \left(\int_0^t f^i(s)\,ds \right) e(f). \qquad (2.6)$$

In the remaining case where $\alpha = \beta = 0$ we define Λ_0^0 by

$$\Lambda_0^0(t)e(f) = te(f). \qquad (2.7)$$

Note that, for all integers α, β with $0 \leq \alpha, \beta \leq N$, the processes Λ_β^α are defined on the whole of $\mathcal{E}$. The processes Λ_j^i defined in (2.4) are called the *N-dimensional gauge processes*. In the case $N = 1$ the sole gauge process Λ_1^1 is sometimes denoted Λ. The processes Λ_j^0 defined in (2.5) are called the *N-dimensional creation processes* and are sometimes also denoted $A_i^\dagger$ or, in the case $N = 1$, simply by $A^\dagger$. The processes Λ_0^i of (2.6) are called the *N-dimensional annihilation processes* and are sometimes also denoted A^i or, in the case $N = 1$, simply by A. The process Λ_0^0 of (2.7) is called the *time process* and is sometimes also denoted by T. We remark that for $t \geq 0$ we have that $T(t) = t\,Id$ where Id denotes the identity operator in $\Gamma(L^2(\mathbf{R}_+; \mathbf{C}^N))$. We will also wish to make use of the *identity process* which takes the value Id for all times t. The identity process is denoted Id and is clearly exponential, adapted and self-adjoint.

Now that we have defined the Λ_β^α we may define processes of the form Λ_B where B is an arbitrary element of $M_0(N)$. Suppose that B may be written as a sum $\lambda_\alpha^\beta E_\beta^\alpha$ where each λ_α^β is a complex number. Then we define the process Λ_B by

$$\Lambda_B = \lambda_\alpha^\beta \Lambda_\beta^\alpha. \qquad (2.8)$$

It is not difficult to show that for all integers α, β with $0 \leq \alpha, \beta \leq N$ and for all $t \geq 0$ we have

$$(\Lambda_\beta^\alpha(t))^\dagger = \Lambda_\alpha^\beta(t) \qquad (2.9)$$

so that $\Lambda_\beta^{\alpha\dagger} = \Lambda_\alpha^\beta$. It follows from this that if, as before, we have some element B of $M_0(N)$ with

$$B = \sum_{\alpha,\beta=0}^{N} \lambda_\alpha^\beta E_\beta^\alpha$$

then Λ_B is equal to

$$\sum_{\alpha,\beta=0}^{N} \lambda_\alpha^\beta \Lambda_\beta^\alpha.$$

It follows from (2.9) that $(\Lambda_B)^\dagger$ is equal to

$$\sum_{\alpha,\beta=0}^{N} \bar{\lambda}_\alpha^\beta \Lambda_\alpha^\beta. \tag{2.10}$$

We can see from (2.10) that if $\bar{B}^T$ denotes the conjugate transpose of B then $(\Lambda_B)^\dagger = \Lambda_{\bar{B}^T}$.

2.3 Construction of Simple Quantum Stochastic Integrals

In this section we develop the notion of a quantum stochastic integral. The notion of quantum stochastic integration was first introduced in [HP1]. We begin with an important proposition relating to the processes Λ_B of the previous section.

Proposition 2.6 For arbitrary $B \in M_0(N)$ and arbitrary $s, t \in \mathbf{R}_+$ with $t < s$ we have that

$$\Lambda_B(s) - \Lambda_B(t) = 1_t \otimes \Lambda_B{}^t(s)$$

where 1_t is the identity operator in $\mathcal{E}([0,t])$ and $\Lambda_B{}^t(s)$ is an operator in $\mathcal{E}((t,\infty))$.

As might be expected, we begin by considering processes that are as uncomplicated as possible. We say that an adapted process F is *elementary* if, for some t_1, t_2 in $\mathbf{R}_+$ with $0 \leq t_1 < t_2 < \infty$ we have

$$F(t) = \begin{cases} 0 & \text{if } 0 \leq t < t_1; \\ F(t_1) & \text{if } t_1 \leq t < t_2; \\ 0 & \text{if } t \geq t_2. \end{cases} \tag{2.11}$$

Equivalently we may write $F = \chi_{[t_1,t_2)} F(t_1)$. We say that a process is *simple* if it is a finite complex linear combination of elementary processes. Linearity allows us to restrict our attention to integrals formed against a single process Λ_β^α for some integers α, β with $0 \leq \alpha, \beta \leq N$.

It is clear that integrating a constant should multiply the constant by the increment in the integrator over the interval of integration. It therefore makes sense to define the integral of an elementary process $F = \chi_{[t_1,t_2)} F(t_1)$ against a process Λ_β^α over the interval $[0,t]$ as follows:

$$\int_0^t F(s) \otimes d\Lambda_\beta^\alpha(s) = \begin{cases} 0 & \text{if } 0 \leq t < t_1; \\ F(t_1)(\Lambda_\beta^\alpha(t) - \Lambda_\beta^\alpha(t_1)) & \text{if } t_1 \leq t < t_2; \\ F(t_1)(\Lambda_\beta^\alpha(t_2) - \Lambda_\beta^\alpha(t_1)) & \text{if } t \geq t_2. \end{cases} \tag{2.12}$$

The reason for the inclusion of the tensor product in the integral on the left of (2.12) will become clear in due course. For s, t in $\mathbf{R}_+$ we denote the quantity $\min\{s, t\}$ by $s \wedge t$. This enables us to re-write (2.12) as

$$\int_0^t F(s) \otimes d\Lambda_\beta^\alpha(s) = F(t_1)(\Lambda_\beta^\alpha(t \wedge t_2) - \Lambda_\beta^\alpha(t \wedge t_1)). \qquad (2.13)$$

Careless inspection of (2.13) arouses the suspicion that illicit multiplication of unbounded operators may be taking place. Although we are indeed composing unbounded operators, this composition is not an illicit one. The operator $F(t_1)$ acts on $\mathcal{E}$ but, as F is assumed to be an adapted process, we have that $F(t_1)$ is of the form $F_{t_1} \otimes 1^{t_1}$. Here F_{t_1} is an operator acting in $\mathcal{E}([0, t_1])$ and 1^{t_1} is the identity operator in $\mathcal{E}((t, \infty))$. On the other hand, for $t \leq t_1$ we have that the increment $\Lambda_\beta^\alpha(t \wedge t_2) - \Lambda_\beta^\alpha(t \wedge t_1)$ is equal to 0. By proposition 2.6 we have that for $t > t_1$

$$\Lambda_\beta^\alpha(t \wedge t_2) - \Lambda_\beta^\alpha(t \wedge t_1) = 1_{t_1} \otimes \Lambda_{\beta\,t}^{\alpha\,t_1} \otimes 1^t$$

where 1_{t_1} is the identity operator in $\mathcal{E}([0, t_1])$, 1^t is the identity operator in $\mathcal{E}((t, \infty))$ and $\Lambda_{\beta\,t}^{\alpha\,t_1}$ is an operator in $\mathcal{E}((t_1, t])$. If $t > t_2$ then $\Lambda_{\beta\,t}^{\alpha\,t_1}$ may in fact be written in the the form $\Lambda_{\beta\,t_2}^{\alpha\,t_1} \otimes 1_t^{t_2}$ with the notation interpreted in the obvious way. This observation enables the composition of operators

$$F(t_1)(\Lambda_\beta^\alpha(t \wedge t_2) - \Lambda_\beta^\alpha(t \wedge t_1))$$

to be re-written for $t \geq t_1$ as

$$F_{t_1} \otimes 1^{t_1}(1_{t_1} \otimes \Lambda_{\beta\,t \wedge t_2}^{\alpha\,t_1} \otimes 1^{t \wedge t_2}).$$

It is clear that in this composition of operators unbounded operators are only composed with the identity. This is perfectly legitimate and for $t \geq t_1$ yields

$$F_{t_1} \otimes \Lambda_{\beta\,t \wedge t_2}^{\alpha\,t_1} \otimes 1^{t \wedge t_2}.$$

Thus (2.13) makes good sense and the reason for the presence of the tensor product in (2.12) becomes clear.

We now take F to be a simple integrand with $F = \sum_{j=1}^n F_j$ where n is some positive integer and each F_j is an elementary process. In this case it is clear that we should define

$$\int_0^t F(s) \otimes d\Lambda_\beta^\alpha(s) = \sum_{j=1}^n \int_0^t F_j(s) \otimes d\Lambda_\beta^\alpha(s). \qquad (2.14)$$

The integral in (2.14) is independent of the choice of decomposition of F and defines an adapted process I for which the adjoint process is defined by

$$I^\dagger(t) = \left(\int_0^t F(s) \otimes d\Lambda_\beta^\alpha(s) \right)^\dagger = \int_0^t F^\dagger(s) \otimes d\Lambda_\alpha^\beta(s).$$

There are two equalities relating to the quantum stochastic integrals just defined that play a major role in this work. These are known as the first and second fundamental formulas and were first presented in [HP1]. We present these in the form of two theorems.

Theorem 2.7 (The first fundamental formula) For arbitrary elements f, g of $L^2(\mathbf{R}_+; \mathbf{C}^N)$, arbitrary $t \geq 0$, arbitrary α, β with $0 \leq \alpha, \beta \leq N$ and an arbitrary simple process F the following equality holds:

$$\langle e(f), \int_0^t F(s) \otimes d\Lambda_\beta^\alpha(s) e(g) \rangle = \int_0^t f_\beta(s) g^\alpha(s) \langle e(f), F(s) e(g) \rangle \, ds.$$

Theorem 2.8 (The second fundamental formula) For arbitrary elements f, g of $L^2(\mathbf{R}_+; \mathbf{C}^N)$, arbitrary $t \geq 0$, arbitrary integers $\alpha, \beta, \gamma, \delta$ with $0 \leq \alpha, \beta, \gamma, \delta \leq N$ and arbitrary simple processes F, H we have

$$\langle \int_0^t F(s) \otimes d\Lambda_\beta^\alpha(s) e(f), \int_0^t H(s) \otimes d\Lambda_\delta^\gamma(s) e(g) \rangle$$

$$= \int_0^t f_\delta(s) g^\gamma(s) \langle \int_0^s F(u) \otimes d\Lambda_\beta^\alpha(u) e(f), H(u) e(g) \rangle \, ds$$

$$+ \int_0^t f_\alpha(s) g^\beta(s) \langle F(s) e(f), \int_0^s H(u) \otimes d\Lambda_\delta^\gamma(u) e(g) \rangle \, ds$$

$$+ \hat{\delta}_\delta^\beta \int_0^t f_\alpha(s) g^\gamma(s) \langle F(s) e(f), H(s) e(g) \rangle \, ds$$

where $\hat{\delta}_\cdot^\cdot$ is as defined in (2.3).

If we disregard the problem of multiplying unbounded operators, the second fundamental formula can be thought of as expressing the following formal relation:

$$\int_0^t F(s) \, d\Lambda_\beta^\alpha(s) \int_0^t H(s) \, d\Lambda_\delta^\gamma(s)$$

$$= \int_0^t F(s) \int_0^s H(u) \, d\Lambda_\delta^\gamma(u) \, d\Lambda_\beta^\alpha(s) + \int_0^t \int_0^s F(s) \, d\Lambda_\beta^\alpha(u) H(s) \, d\Lambda_\delta^\gamma(s)$$

$$+ \hat{\delta}_\delta^\alpha \int_0^t F(s) H(s) \, d\Lambda_\beta^\gamma(s). \tag{2.15}$$

The third term of (2.15) is known as the *Ito correction term* and is characteristic of stochastic calculus. If we set $I_1 = \int F \, d\Lambda_\beta^\alpha$ and $I_2 = \int G \, d\Lambda_\delta^\gamma$ then we may express (2.15) in a formal differential manner as

$$d(I_1 I_2) = I_1 \, dI_2 + (dI_1) I_2 + dI_1 . dI_2. \tag{2.16}$$

Here we have that

$$dI_1.dI_2 = (F \otimes d\Lambda_\beta^\alpha)(H \otimes d\Lambda_\delta^\gamma) = FH \otimes \hat{\delta}_\delta^\alpha d\Lambda_\beta^\gamma.$$

Later on in this work a tensor product will be presented which is defined on $\mathbf{Z}_2$-graded structures. This will yield equalities similar to (2.16) but will also introduce a 'grading factor' which is a power of -1. We see from this presentation that we have a notion of multiplication of differentials. Explicitly we have that

$$d\Lambda_\beta^\alpha.d\Lambda_\delta^\gamma = \hat{\delta}_\delta^\alpha d\Lambda_\beta^\gamma. \tag{2.17}$$

In the case where $N = 1$, relation (2.17) may be expressed as a table:

Table 1 Multiplication of quantum stochastic differentials

.	$dA^\dagger$	$d\Lambda$	dA	dT
$dA^\dagger$	0	0	0	0
$d\Lambda$	$dA^\dagger$	$d\Lambda$	0	0
dA	dT	dA	0	0
dT	0	0	0	0

2.4 Extending the Integral

We must now formulate a notion of the quantum stochastic integral of a process that is not simple. Of particular interest in this work is the case where the integrand itself is a quantum stochastic integral. In order to make this extension to integrals that are not simple we require the fundamental estimate which was first presented in [HP1]. We give this in the form of a theorem.

Theorem 2.9 (Fundamental estimate) Take a family of arbitrary simple processes F_α^β with $0 \le \alpha, \beta \le N$. Let f be an arbitrary element of $L^2(\mathbf{R}_+; \mathbf{C}^N)$ and t be an arbitrary time with $t \ge 0$. Then there exists a constant c_f depending on f with $c_f > 0$ such that for all $s \le t$ we have

$$\left\| \int_0^s F_\alpha^\beta(u) \otimes d\Lambda_\beta^\alpha(u)e(f) \right\|^2 \le c_f \max_{\alpha,\beta} \int_0^t \|F_\alpha^\beta(u)e(f)\|^2 \, du.$$

We say that a process F is *locally simple* if, for each $t \in \mathbf{R}_+$, the process $\chi_{[0,t]}F$ is simple. Note that $\chi_{[0,t]}F$ coincides with F up to time t and vanishes thereafter.

We may extend the notion of quantum stochastic integration from integrands that are simple processes to integrands that are locally simple processes by defining the integral $\int_0^t F(s) \otimes d\Lambda_\beta^\alpha(s)$ of a locally simple process F to be

$$\int_0^t \chi_{[0,t]} F(s) \otimes d\Lambda_\beta^\alpha(s).$$

It is clear that the first and second fundamental formulas (theorems 2.7 and 2.8 respectively) and the fundamental estimate (theorem 2.9) will continue to hold for integrals of locally simple processes.

We say that an adapted process F is *measurable* if, for arbitrary f, g in $L^2(\mathbf{R}_+; \mathbf{C}^N)$, the complex valued function $t \mapsto \langle e(f), F(t)e(g) \rangle$ is Borel measurable. We say that a measurable adapted process F is *integrable* if there exists a sequence $(F_n)_{n \in \mathbf{N}}$ of locally simple adapted processes such that for all $t \in \mathbf{R}_+$ and all $f \in L^2(\mathbf{R}_+; \mathbf{C}^N)$ we have

$$\lim_{n \to \infty} \int_0^t \|(F(s) - F_n(s)e(f)\|^2 \, ds = 0,$$

$$\lim_{n \to \infty} \int_0^t \|(F^\dagger(s) - F_n^\dagger(s)e(f)\|^2 \, ds = 0.$$

(2.18)

We now give a proposition justifying the extension of quantum stochastic integrals to arbitrary integrable processes. As might be expected, the proof of this proposition relies on the fundamental estimate given in theorem 2.9.

Proposition 2.10 Let F be an arbitrary integrable process and α, β be arbitrary integers with $0 \le \alpha, \beta \le N$. Let $(F_n)_{n \in \mathbf{N}}$ be a sequence of locally simple processes approximating F in the sense of (2.18). Define for arbitrary $f \in L^2(\mathbf{R}_+; \mathbf{C}^N)$, arbitrary $t \ge 0$ and arbitrary $n \in \mathbf{N}$ the vectors $\psi_n(f, t)$ and $\psi_n^\dagger(f, t)$ as follows:

$$\psi_n(f, t) := \int_0^t F_n(s) \otimes d\Lambda_\beta^\alpha(s)e(f), \qquad \psi_n^\dagger(f, t) := \int_0^t F_n^\dagger(s) \otimes d\Lambda_\alpha^\beta(s)e(f).$$

Then we have that the sequences of vectors $(\psi_n(f, t))_{n \in \mathbf{N}}$ and $(\psi_n^\dagger(f, t))_{n \in \mathbf{N}}$ are convergent in $\Gamma(L^2(\mathbf{R}_+; \mathbf{C}^N))$. Moreover,

a) The limits $\psi(f, t)$ and $\psi^\dagger(f, t)$ are independent of the choice of approximating sequence $(F_n)_{n \in \mathbf{N}}$.

b) The convergence of each sequence $(\psi_n(f, t))_{n \in \mathbf{N}}$ to $\psi(f, t)$ is uniform in t on each finite interval $[0, s]$ of $\mathbf{R}_+$. The same is true of the convergence of each sequence $(\psi_n^\dagger(f, t))_{n \in \mathbf{N}}$ to $\psi^\dagger(f, t)$.

2.5 Iterated Integrals

In this section we show that quantum stochastic integrals are themselves integrable processes and so may be integrated to yield iterated quantum stochastic integrals. We begin with a definition and then give a series of propositions which lead to the desired conclusion.

Definition 2.11 Take arbitrary integers α, β with $0 \leq \alpha, \beta \leq N$ and an arbitray integrable process F with approximating sequence of simple processes $(F_n)_{n \in \mathbb{N}}$. We define the *stochastic integral process*

$$I = \int_0^{\cdot} F(s) \otimes d\Lambda_{\beta}^{\alpha}(s)$$

by its action on an arbitrary exponential vector $e(f)$ with $f \in L^2(\mathbb{R}_+; \mathbb{C}^N)$ at an arbitrary time $t \geq 0$ as follows:

$$I(t)e(f) = \int_0^t F(s) \otimes d\Lambda_{\beta}^{\alpha}(s)e(f) = \lim_{n \to \infty} \int_0^t F_n(s) \otimes d\Lambda_{\beta}^{\alpha}(s)e(f).$$

Proposition 2.12 For an arbitrary integrable process F and arbitrary integers α, β with $0 \leq \alpha, \beta \leq N$ we have that the process $\int_0^{\cdot} F(s) \otimes d\Lambda_{\beta}^{\alpha}(s)$ is an adapted process. Furthermore, we have for each $t \geq 0$ that

$$\left(\int_0^t F(s) \otimes d\Lambda_{\beta}^{\alpha}(s) \right)^{\dagger} = \int_0^t F^{\dagger}(s) \otimes d\Lambda_{\alpha}^{\beta}(s).$$

This proposition provides us with the motivation to define a map $\dagger$ on the quantum stochastic differentials $d\Lambda_A$ by linear extension of the rule $(d\Lambda_{\beta}^{\alpha})^{\dagger} = d\Lambda_{\alpha}^{\beta}$. It is a simple matter to show that $\dagger$ is an involution on these differentials with respect to the multiplication given by (2.17). This associative algebra with involution structure of $\{d\Lambda_A : A \in M_0(N)\}$ will be exploited in Chapter 5.

Proposition 2.13 The first fundamental formula (theorem 2.7), the second fundamental formula (theorem 2.8) and the fundamental estimate (theorem 2.9) all hold for arbitrary integral processes.

Definition 2.14 We say that an adapted process E is *continuous* if, for each ψ in $\mathcal{E}$, the vector-valued functions

$$t \mapsto E(t)\psi, \quad t \mapsto E^{\dagger}(t)\psi \tag{2.19}$$

are both continuous.

Proposition 2.15 Continuous processes are integrable.

Proof. Define simple processes E_n for $n \in \mathbf{N}$ to approximate the continuous process E by

$$E_n(t) = \begin{cases} E(\frac{i}{n}) & \text{if } \frac{i}{n} \leq t < \frac{i+1}{n} \text{ with } j = 0, \ldots, n^2 - 1; \\ 0 & \text{if } t \geq n. \end{cases}$$

For each $f \in L^2(\mathbf{R}_+; \mathbf{C}^N)$ and each $t \in \mathbf{R}_+$, the function $s \mapsto E(s)e(f)$ is continuous and hence uniformly continuous on $[0, t]$. Suppose we are given some $\epsilon > 0$. The uniform continuity enables us to select $\delta > 0$ such that for arbitrary $s_1, s_2 \in [0, t]$ that satisfy $|s_1 - s_2| < \delta$ we have

$$\|(E(s_1) - E(s_2))e(f)\| < \epsilon^{\frac{1}{2}}t^{-\frac{1}{2}}, \quad \|(E^\dagger(s_1) - E^\dagger(s_2))e(f)\| < \epsilon^{\frac{1}{2}}t^{-\frac{1}{2}}.$$

Choose $n_0 \in \mathbf{N}$ such that $1/n_0 < \delta$. For each $n > n_0$ and each $s \in [0, t]$ denote by $j_{(s,n)}$ the integer such that $\frac{j_{(s,n)}}{n} \leq s < \frac{j_{(s,n)}+1}{n}$. Then we have that

$$\int_0^t \|(E(s) - E_n(s))e(f)\|^2 \, ds =$$

$$\int_0^t \| \left(E(s) - E\left(\frac{j_{(s,n)}}{n}\right)\right) e(f)\|^2 \, ds < \int_0^t \frac{\epsilon}{t} \, ds = \epsilon,$$

$$\int_0^t \|(E^\dagger(s) - E_n{}^\dagger(s))e(f)\|^2 \, ds =$$

$$\int_0^t \| \left(E^\dagger(s) - E^\dagger\left(\frac{j_{(s,n)}}{n}\right)\right) e(f)\|^2 \, ds < \int_0^t \frac{\epsilon}{t} \, ds = \epsilon.$$

It follows from the definition given in (2.18) that E is integrable. ∎

Proposition 2.16 Quantum stochastic integrals are continuous with respect to time.

Proof. Let $I(t) = \int_0^t F(s) \otimes d\Lambda_\beta^\alpha(s)$. If t is a fixed time with $t \geq 0$ and t_1, t_2 are elements of $\mathbf{R}_+$ with $t_1 < t_2 < t$ then we have

$$I(t_2) - I(t_1) = \int_0^t \tilde{F}(s) \otimes d\Lambda_\beta^\alpha(s)$$

where $\tilde{F} = \chi_{[t_1,t_2]}F$. By the fundamental estimate (theorem 2.9), for $f \in L^2(\mathbf{R}_+; \mathbf{C}^N)$ there exists a constant $c_f > 0$ depending on t and f but not on t_1 or t_2 such that

$$\|(I(t_2) - I(t_1))e(f)\|^2 \leq c_f \int_0^t \|\tilde{F}(s)e(f)\|^2 \, ds = c_f \int_{t_1}^{t_2} \|F(s)e(f)\|^2 \, ds.$$

$$(2.20)$$

The right-hand side of (2.20) evidently converges to 0 as $|t_2 - t_1| \to 0$. A similar argument for $I(t)^\dagger$ completes the proof.∎

Corollary 2.17 (To propositions 2.15 and 2.16) Stochastic integral processes are themselves integrable processes.

It follows from corollary 2.17 that for an arbitrary integrable process F we may take iterated integrals of the form

$$\int_{0 < t_1 < \cdots < t_n < t} F(t_1) \, d\Lambda_{\beta_1}^{\alpha_1}(t_1) \cdots d\Lambda_{\beta_n}^{\alpha_n}(t_n).$$

The later chapters of this work are largely concerned with iterated integrals in a graded version of the calculus that we have just described. The one-dimensional precursor of this theory is outlined briefly in the next section.

Note that, in contrast to the adapted integrands of quantum stochastic calculus, the classical (commutative) theory of stochastic calculus includes the Hitsuda-Skorohod integral which may take an anticipating integrand. Furthermore, creation operators are relevant in this branch of the classical theory.

2.6 Boson–Fermion Unification

In this section we outline the Boson–Fermion unification theory that was introduced in [HP2]. This theory applies in the case where the dimension parameter N is equal to 1 so that the Fock space in use is $\Gamma(L^2(\mathbf{R}_+; \mathbf{C}))$.

The *parity process* $R = (R(t))_{t \in \mathbf{R}_+}$ is defined on all of $\Gamma(L^2(\mathbf{R}_+; \mathbf{C}))$ by its action on an arbitrary exponential vector $e(f)$ with $f \in L^2(\mathbf{R}_+; \mathbf{C})$ as follows:

$$R(t)e(f) = e(-\chi_{[0,t]}f + \chi_{(t,\infty)}f). \tag{2.21}$$

If the process Λ is expressed in the form of its spectral decomposition as

$$\Lambda(t) = \sum_{n \in \mathbf{N}} n E_n(t)$$

then the process $(-1)^\Lambda := \sum_{n \in \mathbf{N}} (-1)^n E_n(\cdot)$ is equal to R as defined in (2.21).

The processes $A^\dagger$ and A of one-dimensional quantum stochastic calculus are known properly as the *bosonic* creation and annihilation operators respectively. This is because they are derived ultimately from the canonical commutation relations (CCR) of quantum theory. Given an arbitrary Hilbert space h, a representation of the canonical commutation relations (RCCR) is a family of operators $(a(u), u \in h)$ in a Hilbert space $\mathcal{H}$ satisfying the following relations for all u, v in h and all α in $\mathbf{C}$:

1) $a(u + \alpha v) = a(u) + \bar{\alpha} a(v)$
2) $[a(u), a^\dagger(v)] = \langle u, v \rangle 1$
3) $[a(u), a(v)] = 0$
4) $[a^\dagger(u), a^\dagger(v)] = 0$

where $a^\dagger(w)$ denotes the adjoint of $a(w)$ for arbitrary w in h. If the carrier space $\mathcal{H}$ of an RCCR contains an element Ω satisfying

5) $a(u)\Omega = 0$ for all $u \in h$
6) $\{a^\dagger(u_n) \cdots a^\dagger(u_1)\Omega : n \in \mathbf{N}, \ u_1, \ldots u_n \in h\}$ is total in $\mathcal{H}$

then Ω is said to be a *cyclic vacuum vector* for the RCCR.

If $\mathcal{H}$ is taken to be the Fock space $\Gamma(h)$ and we define $a(u)$ on $\Gamma(h)$ by its action on the exponential vectors $e(f)$ of $\Gamma(h)$ thus:

$$a(u)e(f) = \langle u, f \rangle e(f) \qquad (2.22)$$

then we have an RCCR with cyclic vacuum vector $e(0)$. Any other RCCR with a vacuum vector is unitarily equivalent to this Fock representation. When h is infinite dimensional there exist RCCRs that are inequivalent to the Fock representation but it is only the Fock representation (and equivalent representations thereof) that possesses the cyclic vacuum vector.

The bosonic quantum stochastic calculus introduced in [HP1] was constructed around the CCR. At time t with $t \geq 0$ the annihilation process A takes the value $A(t)$ which is equal to $a(\chi_{[0,t]})$ with a as defined in (2.22) and with $h = L^2(\mathbf{R}_+; \mathbf{C})$.

In [BSW] and [AH] we see a quantum stochastic calculus developed from the canonical *anti*commutation relations (CAR). These are expressed as follows, with u, v being arbitrary elements of a Hilbert space h and α an arbitrary element of $\mathbf{C}$:

1') $b(u)b^\dagger(v) + b^\dagger(v)b(u) = \langle u, v \rangle 1$
2') $b(u)b(v) + b(v)b(u) = 0$
3') $b^\dagger(u)b^\dagger(v) + b^\dagger(v)b^\dagger(u) = 0.$

Identical conclusions relating to equivalence of representations and the vacuum vector hold in the CAR case. The quantum stochastic process B of [AH] is defined at time t with $t \geq 0$ by $B(t) = b(\chi_{[0,t]})$ where again we take the Hilbert space h to be $L^2(\mathbf{R}_+; \mathbf{C})$.

In [HP2] it was shown that the CCR and CAR versions of quantum stochastic calculus can both be realised in one and the same Fock space. This equivalence is based on the following four differential equalities, the latter three all being direct consequences of the first:

$$dA = (-1)^\Lambda \, dB \qquad dB = (-1)^\Lambda \, dA$$
$$dA^\dagger = (-1)^\Lambda \, dB^\dagger \qquad dB^\dagger = (-1)^\Lambda \, dA^\dagger.$$

The $\mathbf{Z}_2$-graded multidimensional quantum stochastic calculus that will be developed in this work is a generalisation of the Boson–Fermion equivalence outlined in this section.

3 Z₂-Graded Structures

3.1 Z₂-Graded Vector Spaces

We denote by $\mathbf{Z}_2$ the abelian group that is the set $\{0, 1\}$ equipped with the standard modulo 2 addition given in the following table:

Table 2 The abeblian group $\mathbf{Z}_2$

+	0	1
0	0	1
1	1	0

When applied to elements of $\mathbf{Z}_2$, the symbol '+' always denotes addition modulo 2. This will not normally be mentioned explicitly.

If V is a vector space over $\mathbf{C}$ then a $\mathbf{Z}_2$-grading of V is a decomposition of V into an internal direct sum $V_0 + V_1$ where V_0 and V_1 are subspaces of V. A vector space equipped with a $\mathbf{Z}_2$-grading is said to be a *$\mathbf{Z}_2$-graded vector space*. Every element v of V may be decomposed uniquely as $v = v_0 + v_1$ where $v_0 \in V_0$ and $v_1 \in V_1$. The elements v_0 and v_1 are known as the *homogeneous components* of v. The elements of $V_0 \cup V_1$ are said to be *homogeneous, of definite parity* or *of definite degree*. The elements of V_0 are said to be *even, of parity* 0 or *of degree* 0. The elements of V_1 are said to be *odd, of parity* 1 or *of degree* 1. The zero element 0 of V is the unique element of $V_0 \cap V_1$ and so is both odd and even. Throughout this work, for any $\mathbf{Z}_2$-graded vector space or any structure with an underlying $\mathbf{Z}_2$-graded vector space (such as the $\mathbf{Z}_2$-graded algebras to be defined in the next section) we denote the parity of a homogeneous element v of of V by $\sigma(v)$. We do not use distinct symbols for the parity function σ in distinct $\mathbf{Z}_2$-graded structures. Note that σ is not well-defined on $0 \in V$. In practice we take $\sigma(0)$ to be 0 or 1 as is convenient.

Consider a subspace U of V. If each element u of U has its homogeneous components in U then we say that U is a *$\mathbf{Z}_2$-graded subspace* of V. If V and W are both $\mathbf{Z}_2$-graded vector spaces then a linear mapping $g: V \to W$ is said to be *homogeneous of degree α* where α in an element of $\mathbf{Z}_2$ if

$$g(V_\beta) \subset W_{\beta+\alpha} \qquad \text{for all } \beta \in \mathbf{Z}_2.$$

The mapping g is called a $\mathbf{Z}_2$-*graded vector space homomorphism* from V to W if g is a homogeneous linear mapping of degree 0. A $\mathbf{Z}_2$-*graded vector space isomorphism* is defined in the obvious way as a bijective $\mathbf{Z}_2$-graded vector space homomorphism.

If $V = V_0 + V_1$ and $V' = V_0' + V_1'$ are two $\mathbf{Z}_2$-graded vector spaces over $\mathbf{C}$ then the tensor product $V \otimes V'$ has a canonical $\mathbf{Z}_2$-grading $(V \otimes V')_0 + (V \otimes V')_1$ where

$$(V \otimes V')_0 = V_0 \otimes V_0' + V_1 \otimes V_1', \qquad (V \otimes V')_1 = V_0 \otimes V_1' + V_0 \otimes V_1'.$$

If W and W' are a further pair of $\mathbf{Z}_2$-graded vector spaces and $g: V \to W$ and $g': V' \to W'$ are a pair of homogeneous linear mappings with respective degrees $\sigma(g)$ and $\sigma(g')$ then we define the linear mapping

$$g \otimes g': V \otimes V' \to W \otimes W'$$

by

$$(g \otimes g')(v \otimes v') = (-1)^{\sigma(v)\sigma(g')} g(v) \otimes g(v').$$

It is clear that $g \otimes g'$ is a homogeneous linear mapping of parity $\sigma(g) + \sigma(g')$. It is easy to see how these definitions can be extended to tensor products of finitely many $\mathbf{Z}_2$-graded vector spaces.

3.2 $\mathbf{Z}_2$-Graded Algebras

Let A be an algebra (not necessarily associative or possessing a unit) over $\mathbf{C}$. We say that the algebra A is $\mathbf{Z}_2$-graded if the underlying vector space of A is $\mathbf{Z}_2$-graded, that is to say, there is a decomposition $A_0 + A_1$ of A as a vector space where A_0 is the even subspace of A and A_1 is the odd subspace of A and, furthermore, for all α, β in $\mathbf{Z}_2$ we have

$$A_\alpha A_\beta \subset A_{\alpha+\beta}. \tag{3.1}$$

It is common for $\mathbf{Z}_2$-graded algebras to be called *superalgebras*.

Proposition 3.1 If a $\mathbf{Z}_2$-graded algebra A has a unit element 1 then 1 is even.

Proof. We have for all a in A that $1a = a1 = a$. Suppose 1 has homogeneous decomposition $1 = 1_0 + 1_1$. Then for all a_0 in the even subspace A_0 of A we have

$$1a_0 = (1_0 + 1_1)a_0 = 1_0 a_0 + 1_1 a_0 = a_0$$
$$a_0 1 = a_0(1_0 + 1_1) = a_0 1_0 + a_0 1_1 = a_0.$$

As a_0 is even, by condition (3.1) we must have $1_1 a_0 = a_0 1_1 = 0$ for all even a_0 in A. Similarly, for all a_1 in the odd subspace A_1 of A we have

$$1a_1 = (1_0 + 1_1)a_1 = 1_0a_1 + 1_1a_1 = a_1$$
$$a_11 = a_1(1_0 + 1_1) = a_11_0 + a_11_1 = a_1.$$

As a_1 is odd, by condition (3.1) we must have $a_11_1 = 1_1a_1 = 0$ for all odd a_1 in A. We know that an arbitrary element of A can be decomposed uniquely as $a = a_0 + a_1$ with a_0 even and a_1 odd. Thus we have that for all $a \in A$

$$a = 1a = 1(a_0 + a_1) = 1_0a_0 + 1_0a_1 = 1_0(a_0 + a_1) = 1_0a$$
$$a = a1 = (a_0 + a_1)1 = a_01_0 + a_11_0 = (a_0 + a_1)1_0 = a1_0$$

so that $1 = 1_0$ and we may conclude that 1 is an even element of A.∎

A *homomorphism of $\mathbf{Z}_2$-graded algebras* is defined to be a linear homogeneous map that is a homomorphism of the underlying algebras and furthermore is of degree 0, i.e., is grade preserving. Thus we have that, if $\theta: A \to B$ is a homomorphism of $\mathbf{Z}_2$-graded algebras, then for arbitrary a, a' in A

$$\theta(aa') = \theta(a)\theta(a')$$
$$\theta(a + a') = \theta(a) + \theta(a')$$
$$\sigma(\theta(a)) = \sigma(a) \text{ when } a \text{ is homogeneous.}$$

An isomorphism of $\mathbf{Z}_2$-graded algebras is, of course, a bijective homomorphism of $\mathbf{Z}_2$-graded algebras.

A $\mathbf{Z}_2$-graded *subalgebra* (or subsuperalgebra) of a $\mathbf{Z}_2$-graded algebra A is a subalgebra B of the underlying algebra of A such that the underlying vector space of B is a $\mathbf{Z}_2$-graded sub-vector space of the underlying $\mathbf{Z}_2$-graded vector space of A. A $\mathbf{Z}_2$-graded graded ideal of a $\mathbf{Z}_2$-graded algebra A is an ideal K of the underlying algebra of A such that the underlying vector space of K is a $\mathbf{Z}_2$-graded sub-vector space of the underlying $\mathbf{Z}_2$-graded vector space of A.

Proposition 3.2 If A is a $\mathbf{Z}_2$-graded algebra and H is a two-sided $\mathbf{Z}_2$-graded ideal of A then the quotient A/H is a $\mathbf{Z}_2$-graded algebra.

Proof. Consider an arbitrary element aH of A/H. As both A and H are $\mathbf{Z}_2$-graded, a may be decomposed as $a_0 + a_1$ and H decomposed as $H_0 + H_1$. Therefore the coset aH is decomposed into even and odd parts as $(a_0H_0 + a_1H_1) + (a_1H_0 + a_0H_1)$. Suppose that $bH = aH$ where b is another element of A. Then $b = ah$ where h is some element of H. The decomposition of bH is

$$(b_0H_0 + b_1H_1) + (b_1H_0 + b_0H_1). \tag{3.2}$$

The element h may be decomposed into its homogeneous components so that we may express h as $h_0 + h_1$. We may now re-write (3.2) as

$$((a_0h_0 + a_1h_1)H_0 + (a_0h_1 + a_1h_0)H_1) + ((a_0h_1 + a_1h_0)H_0 + (a_0h_0 + a_1h_1)H_1)$$

by simple re-arrangement this is equal to

$$a_0(h_0 + h_1)H_0 + a_1(h_0 + h_1)H_1 + a_1(h_0 + h_1)H_0 + a_0(h_0 + h_1)H_1$$

which is equal to

$$(a_0 H_0 + a_1 H_1) + (a_1 H_0 + a_0 H_1)$$

so that the decomposition of the coset cH into its homogeneous components is independent of c and we have a consistent $\mathbf{Z}_2$-grading for A/H.■

If A and A' are both associative $\mathbf{Z}_2$-graded algebras not necessarily possessing a unit then we define a multiplication in the $\mathbf{Z}_2$-graded tensor product space $A \otimes A'$ by linear extension of the following rule for homogeneous a_1, a_1' in A and homogeneous a_2, a_2' in A':

$$(a_1 \otimes a_2)(a_1' \otimes a_2') = (-1)^{\sigma(a_2)\sigma(a_1')} a_1 a_1' \otimes a_2 a_2'. \qquad (3.3)$$

This is known as the *Chevalley tensor product* and is described in full detail in [C]. It is shown in [C] that $A \otimes A'$ under the multiplication given in (3.3) is an associative $\mathbf{Z}_2$-graded algebra. If both A and A' possess units then $A \otimes A'$ possesses a unit in the form of $1_A \otimes 1_{A'}$. Note that the multiplication given in (3.3) is different from the multiplication assigned to tensor products of ungraded associative algebras. It will be necessary in this work to use both types of multiplication. From this point onwards we will normally assume that the graded multiplication of (3.3) is in force in cases where it is not stated otherwise. In principle, it should always be clear which multiplication is in force from the spaces that are being held under consideration.

The multiplication of (3.3) may be extended to n-fold tensor products of associative $\mathbf{Z}_2$-graded algebras. Suppose we are given the associative $\mathbf{Z}_2$-graded algebras (still not necessarily possessing a unit) $A_1, \ldots, A_n$ with $n \geq 1$. We define the multiplication in $A_1 \otimes \cdots \otimes A_n$ by linear extension of the following rule for homogeneous a_i, a_i' in A_i with $i = 1, \ldots, n$:

$$(a_1 \otimes \cdots \otimes a_n)(a_1' \otimes \cdots \otimes a_n') = (-1)^{\sum_{i > j} \sigma(a_i)\sigma(a_j')} a_1 a_1' \otimes \cdots \otimes a_n a_n'.$$

When equipped with this multiplication, $A_1 \otimes \cdots \otimes A_n$ is an associative $\mathbf{Z}_2$-graded algebra. This $\mathbf{Z}_2$-graded algebra will possess a unit $1_{A_1} \otimes \cdots \otimes 1_{A_n}$ if each of the A_i is unital. The space $A_1 \otimes \cdots \otimes A_n$ and its multiplication can be understood in terms of $(\cdots((A_1 \otimes A_2) \otimes A_3) \otimes \cdots \otimes A_{n-1}) \otimes A_n$ using the fact that, if A_1, A_2, A_3 are any associative $\mathbf{Z}_2$-graded algebras, then we have that $A_1 \otimes (A_2 \otimes A_3)$ is isomorphic to $(A_1 \otimes A_2) \otimes A_3$.

Suppose A is an associative $\mathbf{Z}_2$-graded algebra and M is a left A-module in the sense of being a left module with respect to the underlying algebra of A. We say that M is a $\mathbf{Z}_2$-*graded left A-module* if the underlying vector space of M is $\mathbf{Z}_2$-graded and we have for all α, β in $\mathbf{Z}_2$ that

$$A_\alpha \cdot M_\beta \subset M_{\alpha+\beta}$$

where $\cdot$ denotes the action of A on M. It is now clear what the definition of a $\mathbf{Z}_2$-*graded right A-module* should be.

An involution on a $\mathbf{Z}_2$-graded algebra A is a homogeneous conjugate linear map $\dagger \colon A \to A$ such that, for an arbitrary element a of A we have $(a^\dagger)^\dagger = a$ and, if b is a second arbitrary element of A, we have $(ab)^\dagger = b^\dagger a^\dagger$. Note that the antimultiplicity requirement forces the involution to be grade preserving; if a and b are both odd and $\dagger$ were also odd then $(ab)^\dagger$ would be odd whereas $b^\dagger a^\dagger$ would be even.

Take A and B to be $\mathbf{Z}_2$-graded algebras equipped with involutions both of which we denote by $\dagger$. A $\dagger$-*morphism* from A to B is a map $g \colon A \to B$ such that, for each $a \in A$, we have $g(a^\dagger) = g(a)^\dagger$. An involution may be defined on n-fold tensor products of $\mathbf{Z}_2$-graded associative algebras; this is a rather technical matter and will be treated fully in Section 6.1.

3.3 Lie Superalgebras

It is convenient to refer to Lie $\mathbf{Z}_2$-graded algebras as Lie superalgebras and this is the convention adopted in [K,S]. We will begin by defining this entity.

Definition 3.3 A *Lie superalgebra* is a superalgebra L whose multiplication, which we denote $\{.,.\}$, satisfies the following relations for arbitrary homogeneous a, b, c in L:

$$\{a, b\} = -(-1)^{\sigma(a)\sigma(b)}\{b, a\} \tag{3.4}$$

$$\begin{aligned} (-1)^{\sigma(a)\sigma(c)}\{a, \{b, c\}\} \\ +(-1)^{\sigma(b)\sigma(a)}\{b, \{c, a\}\} \\ +(-1)^{\sigma(c)\sigma(b)}\{c, \{a, b\}\} = 0 \end{aligned} \tag{3.5}$$

Identity (3.4) may be described as graded skew-symmetry and (3.5) is known as the graded Jacobi identity.

Note that Lie superalgebras should not be referred to as "$\mathbf{Z}_2$-graded Lie algebras" or "super Lie algebras". Lie superalgebras are not in general Lie algebras. As a Lie superalgebra is simply a particular kind of $\mathbf{Z}_2$-graded algebra, the definitions of a $\mathbf{Z}_2$-graded subalgebra, a $\mathbf{Z}_2$-graded ideal and a $\mathbf{Z}_2$-graded quotient superalgebra in the case of a Lie superalgebra are exactly those described in the previous section. Note also that for all α, β in $\mathbf{Z}_2$ we have $\{L_\alpha, L_\beta\} \subset L_{\alpha+\beta}$ and that all homomorphisms of Lie superalgebras are homogeneous linear maps of degree 0. If $\dagger$ is an involution on a Lie superalgebra L then for arbitrary $a, b \in L$ we have $\{a, b\}^\dagger = \{b^\dagger, a^\dagger\}$. A Lie superalgebra equipped with an involution will be referred to as a Lie $\dagger$-superalgebra.

An important means of constructing Lie superalgebras will now be described. All Lie superalgebras in this work are, directly or indirectly, obtained by this method. Suppose A is an associative superalgebra. Take A_{SLie} to be the underlying $\mathbf{Z}_2$-graded vector space of A equipped with the multiplication $\{.,.\}_{A_{SLie}}$ defined by linear extension of the following rule for homogeneous elements a, b of A:

$$\{a, b\}_{A_{SLie}} = ab - (-1)^{\sigma(a)\sigma(b)}ba. \tag{3.6}$$

The multiplication in force on the right of (3.6) is the one with which the associative superalgebra A is equipped.

We claim that A_{SLie} is a Lie superalgebra. To show this we must demonstrate that (3.4) and (3.5) hold. For (3.4) we see that with arbitrary homogeneous a and b in A we have

$$- (-1)^{\sigma(a)\sigma(b)}\{b, a\}_{A_{SLie}}$$
$$= - (1)^{\sigma(a)\sigma(b)}(ba - (-1)^{\sigma(b)\sigma(a)}ab)$$
$$= - (-1)^{\sigma(a)\sigma(b)}ba + ab$$
$$= ab - (-1)^{\sigma(a)\sigma(b)}ba$$
$$= \{a, b\}_{A_{SLie}}.$$

We show that (3.5) holds for A_{SLie} by straightforward calculation:

$$(-1)^{\sigma(a)\sigma(c)}\{a, \{b, c\}_{A_{SLie}}\}_{A_{SLie}}$$
$$+ (-1)^{\sigma(b)\sigma(a)}\{b, \{c, a\}_{A_{SLie}}\}_{A_{SLie}}$$
$$+ (-1)^{\sigma(c)\sigma(b)}\{c, \{a, b\}_{A_{SLie}}\}_{A_{SLie}}$$
$$= (-1)^{\sigma(a)\sigma(c)}\{a, (bc - (-1)^{\sigma(b)\sigma(c)}cb)\}_{A_{SLie}}$$
$$+ (-1)^{\sigma(b)\sigma(a)}\{b, (ca - (-1)^{\sigma(c)\sigma(a)}ac)\}_{A_{SLie}}$$
$$+ (-1)^{\sigma(c)\sigma(b)}\{c, (ab - (-1)^{\sigma(a)\sigma(b)}ba)\}_{A_{SLie}}$$
$$= (-1)^{\sigma(a)\sigma(c)}abc - (-1)^{\sigma(a)\sigma(c)+\sigma(a)(\sigma(b)+\sigma(c))}bca$$
$$- (-1)^{\sigma(a)\sigma(c)+\sigma(b)\sigma(c)}acb + (-1)^{\sigma(a)\sigma(c)+\sigma(b)\sigma(c)+\sigma(a)(\sigma(c)+\sigma(b))}cba$$
$$+ (-1)^{\sigma(b)\sigma(a)}bca - (-1)^{\sigma(b)\sigma(a)+\sigma(b)(\sigma(c)+\sigma(a))}cab$$
$$- (-1)^{\sigma(b)\sigma(a)+\sigma(c)\sigma(a)}bac + (-1)^{\sigma(b)\sigma(a)+\sigma(c)\sigma(a)+\sigma(b)(\sigma(a)+\sigma(c))}acb$$
$$+ (-1)^{\sigma(c)\sigma(b)}cab - (-1)^{\sigma(c)\sigma(b)+\sigma(c)(\sigma(a)+\sigma(b))}abc$$
$$- (-1)^{\sigma(c)\sigma(b)+\sigma(a)\sigma(b)}cba + (-1)^{\sigma(c)\sigma(b)+\sigma(a)\sigma(b)+\sigma(c)(\sigma(b)+\sigma(a))}bac$$
$$= 0.$$

We now give an example of a class of Lie superalgebras produced by the construction just described. Consider the unital associative algebra $M(N)$ of $N \times N$ complex matrices under standard matrix multiplication. Choose an integer r such that $1 \le r < N$. Define $M(N)_0$ to be the vector subspace of $M(N)$ consisting of all matrices of the form

$$\begin{pmatrix} A_1 & 0 \\ 0 & A_2 \end{pmatrix}$$

where A_1 denotes an arbitrary complex $r \times r$ matrix and A_2 denotes an arbitrary complex $(N-r) \times (N-r)$ matrix. Define $M(N)_1$ to be the vector subspace of $M(N)$ consisting of all matrices of the form

$$\begin{pmatrix} 0 & B_1 \\ B_2 & 0 \end{pmatrix}$$

where B_1 denotes an arbitrary complex $(N-r) \times r$ matrix and B_2 denotes an arbitrary complex $r \times (N-r)$ matrix. It is easy to show that $M(N)$ is equal to the internal direct sum $M(N)_0 + M(N)_1$ and that, for arbitrary α, β in $\mathbf{Z}_2$, we have $M(N)_\alpha M(N)_\beta \subset M(N)_{\alpha+\beta}$. Thus $M(N)$ is a (unital) associative superalgebra and hence is susceptible to the construction described above. The multiplication in the Lie superalgebra $M(N)_{SLie}$ will be denoted $\{.,.\}_{M(N)_{SLie}}$ and defined by linear extension of the the following rule for arbitrary homogeneous A, B in $M(N)_{SLie}$:

$$\{A, B\}_{M(N)_{SLie}} = AB - (-1)^{\sigma(A)\sigma(B)} BA. \tag{3.7}$$

The multiplication on the right of (3.7) is standard matrix multiplication. We may equip $M(N)_{SLie}$ with an involution $\dagger$ by prescribing that, for arbitrary A in $M(N)_{SLie}$, we have $A^\dagger = \bar{A}^T$ where $\bar{A}^T$ denotes the conjugate transpose of A. Elementary linear algebra provides us with the $M(N)$ result that $(\overline{AB})^T = \bar{B}^T \bar{A}^T$ so we may deduce from (3.7) that $\{A, B\}^\dagger = \{B^\dagger, A^\dagger\}$. Thus $M(N)_{SLie}$ is a Lie $\dagger$-superalgebra. A graded version of $M_0(N)$, the variant matrix algebra described in Chapter 2, will prove to be of great importance in the remainder of this work.

From matrices it is natural to move on to more general representations. Take a $\mathbf{Z}_2$-graded vector space $V = V_0 + V_1$ and consider the unital associative algebra $Hom(V)$ of linear mappings in V. This may be $\mathbf{Z}_2$-graded by defining $Hom(V)_0$ and $Hom(V)_1$ by

$$Hom(V)_0 = \{A \in Hom(V) : A(V_\alpha) \subset V_\alpha \text{ for all } \alpha \in \mathbf{Z}_2\}$$
$$Hom(V)_1 = \{A \in Hom(V) : A(V_\alpha) \subset V_{\alpha+1} \text{ for all } \alpha \in \mathbf{Z}_2\}.$$

This is equivalent to defining $Hom(V)_0$ to be the space of homomorphisms in V and defining $Hom(V)_1$ to be the space of all homogeneous linear mappings of degree 1 in V. It is clear that $Hom(V) = Hom(V)_0 + Hom(V)_1$ and that, for all α, β in $\mathbf{Z}_2$, we have $Hom(V)_\alpha Hom(V)_\beta \subset Hom(V)_{\alpha+\beta}$. We now have a unital associative superalgebra structure for $Hom(V)$ and so may construct $Hom(V)_{SLie}$ in the manner described above. The above example of $M(N)_{SLie}$ is $Hom(V)_{SLie}$ in the case where V is a finite dimensional complex vector space graded in the appropriate way.

We may now define the notion of a graded representation.

Definition 3.4 A *graded representation* of a Lie superalgebra L in a $\mathbf{Z}_2$-graded vector space V is a homomorphism of L into $Hom(V)_{SLie}$.

3.4 The Universal Enveloping Superalgebra

The universal enveloping superalgebra of a Lie superalgebra is a construction that will be used a great deal in the latter part of this work. In this section we give the definition and some properties of the universal enveloping superalgebra.

Definition 3.5 Let L be a Lie superalgebra. Consider a pair $(\mathcal{U}, \imath)$ where $\mathcal{U}$ is a unital associative superalgebra and $\imath: L \to \mathcal{U}_{SLie}$ is a homomorphism of Lie superalgebras. Let A be any unital associative superalgebra. Suppose that to any Lie superalgebra homomorphism $\phi: L \to A_{SLie}$ there corresponds a superalgebra homomorphism $\tilde{\phi}: \mathcal{U} \to A$ with $\phi = \tilde{\phi} \circ \imath$. Then $(\mathcal{U}, \imath)$ is the universal enveloping superalgebra of L.

We say that $\tilde{\phi}$ *extends* ϕ. It follows from definition 3.5 that, if $(\mathcal{U}', \imath')$ is a second pair satisfying the conditions stated in definition 3.5 for a particular Lie superalgebra, then there exists an isomorphism $\theta: \mathcal{U} \to \mathcal{U}'$ with $\imath' = \theta \circ \imath$. This is why we may speak of *the* universal enveloping superalgebra of a Lie superalgebra. An explicit construction of such a superalgebra is given in [S]. In practice we consider the elements of L as elements of $\mathcal{U}$, identifying each a in L with $\imath(a) \in \mathcal{U}$. Note that $\mathcal{U}$ always possesses a unit $1_{\mathcal{U}}$ and if $\tilde{\phi}: \mathcal{U} \to A$ extends $\phi: L \to A_{SLie}$ then $\tilde{\phi}(1_{\mathcal{U}}) = 1_A$ where A is any unital associative superalgebra with unit 1_A. Note further that if ϕ possesses the †-morphsim property then its extension $\tilde{\phi}$ will also posses the †-morphism property.

An important theorem concerning the universal enveloping superalgebra of a Lie superalgebra is the $\mathbf{Z}_2$-graded version of the Poincaré-Birkoff-Witt theorem. This theorem gives an explicit canonical basis for $\mathcal{U}$ and will be stated here. Further details and a proof can be found in [S].

Theorem 3.6 ($\mathbf{Z}_2$-graded Poincaré-Birkoff-Witt theorem) Suppose L is a Lie superalgebra with a basis $(L_i)_{i \in X}$ where the L_i are all homogeneous and X is a totally ordered index set. Let $\tilde{X}$ be the set of all finite sequences $(i_1, \ldots, i_s)$ in X so that s is an arbitrary non-negative integer, $i_1 \leq \cdots \leq i_s$ and, whenever the pair of basis elements $L_{i_p}, L_{i_{p+1}}$ are both odd, $i_p < i_{p+1}$. For arbitrary $\alpha = (i_1, \ldots, i_s)$ in $\tilde{X}$ define the element $L_\alpha \in \mathcal{U}$ by $L_\alpha = L_{i_1} \cdots L_{i_s}$. Then the family $(L_\alpha)_{\alpha \in \tilde{X}}$ of elements of $\mathcal{U}$ forms a basis for $\mathcal{U}$.

Continuing with the notation of theorem 3.6, if $\alpha = (i_1, \ldots, i_s)$ is an element of $\tilde{X}$ then we denote the length s of α by $|\alpha|$. For an arbitrary element α of $\tilde{X}$ we define the *degree* of the basis element L_α of $\mathcal{U}$ to be $|\alpha|$.

Any element U of $\mathcal{U}$ will have a component in the basis of $\mathcal{U}$ just described that is of maximal degree. The degree of U is taken to be the degree of this maximal component and is denoted $\deg U$. This notion of degree will be needed in Chapter 8. We finish this chapter with a corollary to theorem 3.6 which will also be needed in Chapter 8.

Corollary 3.7 Let $\mathcal{U}$ be the universal enveloping superalgebra of a Lie superalgebra L. Let $\mathcal{K}$ be the ideal of $\mathcal{U}$ generated by L so that $\mathcal{K} = \mathcal{U}L\mathcal{U}$. Then $\mathcal{U}$ is equal to the internal direct sum $\mathbf{C}1_{\mathcal{U}} + \mathcal{K}$.

4. Representations of Lie Superalgebras in Z_2-Graded Quantum Stochastic Calculus

4.1 Lie Algebra Representations in Ungraded Quantum Stochastic Calculus

As seen in Chapter 2, the integrator processes of N-dimensional quantum stochastic calculus are conveniently denoted as Λ_β^α where α, β are integers with $0 \le \alpha, \beta \le N$. The processes Λ_β^α consist of operators $\Lambda_\beta^\alpha(t)$, $t \ge 0$ acting in Boson Fock space $\Gamma(L^2(\mathbf{R}_+; \mathbf{C}^N))$. The processes Λ_j^i where $1 \le i, j \le N$ form a time indexed family of representations of the complex Lie algebra $gl(N)$ of all $N \times N$ matrices in the 'weak' sense that, for an arbitrary time $t \ge 0$, arbitrary exponential vectors $e(f), e(g)$ with f, g in $L^2(\mathbf{R}_+; \mathbf{C}^N)$ and arbitrary i, j, k, l with $1 \le i, j, k, l \le N$ we have

$$
\begin{aligned}
&\langle \Lambda_j^i(t)^\dagger e(f), \Lambda_l^k(t) e(g) \rangle - \langle \Lambda_l^k(t)^\dagger e(f), \Lambda_j^i(t) e(g) \rangle \\
&= \langle e(f), (\delta_l^i \Lambda_j^k(t) - \delta_j^k \Lambda_l^i(t) e(g) \rangle
\end{aligned}
\tag{4.1}
$$

where δ: is the Kronecker delta with $\delta_n^m = 0$ if $m \ne n$ and $\delta_n^m = 1$ if $m = n$. Equation (4.1) is a rigorous statement of the formal commutation relation

$$
[\Lambda_j^i(t), \Lambda_l^k(t)] = \delta_l^i \Lambda_j^k(t) - \delta_j^k \Lambda_l^i(t).
\tag{4.2}
$$

If A, B are arbitrary elements of $gl(N)$ then (4.2) may be generalised by bilinear extension to the commutation relation

$$
[\Lambda_A, \Lambda_B] = \Lambda_{[A,B]}.
$$

If the time, creation and annihilation fields are to be included then (4.1) must be re-written as

$$
\begin{aligned}
&\langle \Lambda_\beta^\alpha(t)^\dagger e(f), \Lambda_\delta^\gamma(t) e(g) \rangle - \langle \Lambda_\delta^\gamma(t)^\dagger e(f), \Lambda_\beta^\alpha(t) e(g) \rangle \\
&= \langle e(f), (\hat{\delta}_\delta^\alpha \Lambda_\beta^\gamma(t) - \hat{\delta}_\beta^\gamma \Lambda_\delta^\alpha(t) e(g) \rangle
\end{aligned}
\tag{4.3}
$$

where $\alpha, \beta, \gamma, \delta$ are arbitrary integers with $0 \le \alpha, \beta, \gamma, \delta \le N$ and $\hat{\delta}_\beta^\alpha$ is the 'Evans delta' introduced in [Ev] and defined by (2.3) which is 0 unless $\alpha = \beta \ne 0$ in which case it is 1. A formal expression of (4.3) is

$$
[\Lambda_\beta^\alpha(t), \Lambda_\delta^\gamma(t)] = \hat{\delta}_\delta^\alpha \Lambda_\beta^\gamma(t) - \hat{\delta}_\beta^\gamma \Lambda_\delta^\alpha(t).
\tag{4.4}
$$

As in Chapter 2, we denote by $M_0(N)$ the space of all $(N+1) \times (N+1)$ complex matrices indexed from 0 to N. Recall that the basis elements of this space are denoted E_β^α where $0 \le \alpha, \beta \le N$ and E_β^α is the matrix which has zero for each entry except for a 1 in the αth column at the βth row. Let $((\hat\delta))$ be the matrix with $(\alpha, \beta)^{\text{th}}$ entry $\hat\delta_\beta^\alpha$, that is to say

$$((\hat\delta)) = \begin{pmatrix} 0 & 0 & \cdots & 0 \\ 0 & 1 & \cdots & 0 \\ \vdots & \vdots & \ddots & \vdots \\ 0 & 0 & \cdots & 1 \end{pmatrix} = \sum_{i=1}^{N} E_i^i.$$

Recall further that the space $M_0(N)$ is equipped with a multiplication denoted . defined using $((\hat\delta))$ by

$$A.B = A((\hat\delta))B \qquad (4.5)$$

where A, B are arbitrary elements of $M_0(N)$ and the products on the right of (4.5) are standard matrix products.

We may equip $M_0(N)$ with a bracket $[\,.\,,.\,]_{((\hat\delta))}$ defined on arbitrary elements A, B of $M_0(N)$ by

$$[A, B]_{((\hat\delta))} = A.B - B.A = A((\hat\delta))B - B((\hat\delta))A.$$

It is clear that $(M_0(N), .)$ is an associative algebra and so it follows from standard Lie algebra theory [J] that the space $M_0(N)$ equipped with $[\,.\,,.\,]_{((\hat\delta))}$ is a Lie algebra. We shall denote this Lie algebra by $gl_0(N)$. Bilinear extension of (4.4) gives that the Λ_β^α provide a time-indexed family of representations of the Lie algebra $gl_0(N)$, that is to say that the formal relation

$$[\Lambda_A(t), \Lambda_B(t)] = \Lambda_{[A,B]_{((\hat\delta))}}(t)$$

holds for arbitrary $t \ge 0$ and arbitrary $A, B \in gl_0(N)$.

The Lie algebra representation properties of ungraded quantum stochastic calculus were first noted in [HP4].

4.2 Boson–Fermion Equivalence in Quantum Stochastic Calculus of Dimension 1

In [HP2] a Boson–Fermion equivalence was determined for quantum stochastic calculus of dimension $N = 1$. This has already been outlined in Section 2.6. In this section we give a second, slightly more detailed review of this result. This time we shall present the material from the standpoint of the theory that is to follow.

Consider the associative algebra $M_0(1)$ of 2×2 matrices with multiplication defined as in (4.5) by

$$A.B = A \begin{pmatrix} 0 & 0 \\ 0 & 1 \end{pmatrix} B.$$

This associative algebra may be decomposed into the internal direct sum $M_0(1)_0 + M_0(1)_1$ of subspaces defined by

$$M_0(1)_0 = \left\{ \begin{pmatrix} z_{00} & 0 \\ 0 & z_{11} \end{pmatrix} : z_{00}, z_{11} \in \mathbf{C} \right\} = \mathrm{span}\{E_0^0, E_1^1\}$$

$$M_0(1)_1 = \left\{ \begin{pmatrix} 0 & z_{10} \\ z_{01} & 0 \end{pmatrix} : z_{10}, z_{01} \in \mathbf{C} \right\} = \mathrm{span}\{E_1^0, E_0^1\}.$$

Consider the following four matrix multiplication calculations:

$$\begin{pmatrix} z_{00} & 0 \\ 0 & z_{11} \end{pmatrix} \begin{pmatrix} 0 & 0 \\ 0 & 1 \end{pmatrix} \begin{pmatrix} w_{00} & 0 \\ 0 & w_{11} \end{pmatrix} = \begin{pmatrix} 0 & 0 \\ 0 & z_{11}w_{11} \end{pmatrix}$$

$$\begin{pmatrix} 0 & z_{10} \\ z_{01} & 0 \end{pmatrix} \begin{pmatrix} 0 & 0 \\ 0 & 1 \end{pmatrix} \begin{pmatrix} w_{00} & 0 \\ 0 & w_{11} \end{pmatrix} = \begin{pmatrix} 0 & z_{10}w_{11} \\ 0 & 0 \end{pmatrix}$$

$$\begin{pmatrix} z_{00} & 0 \\ 0 & z_{11} \end{pmatrix} \begin{pmatrix} 0 & 0 \\ 0 & 1 \end{pmatrix} \begin{pmatrix} 0 & w_{10} \\ w_{01} & 0 \end{pmatrix} = \begin{pmatrix} 0 & 0 \\ z_{11}w_{01} & 0 \end{pmatrix}$$

$$\begin{pmatrix} 0 & z_{10} \\ z_{01} & 0 \end{pmatrix} \begin{pmatrix} 0 & 0 \\ 0 & 1 \end{pmatrix} \begin{pmatrix} 0 & w_{10} \\ w_{01} & 0 \end{pmatrix} = \begin{pmatrix} z_{10}w_{01} & 0 \\ 0 & 0 \end{pmatrix}.$$

It follows from these simple computations that the following inclusions hold:

$$M_0(1)_0.M_0(1)_0, \; M_0(1)_1.M_0(1)_1 \subset M_0(N)_0;$$
$$M_0(1)_1.M_0(1)_0, \; M_0(1)_0.M_0(1)_1 \subset M_0(N)_1.$$

Consequently, the decomposition of $M_0(1)$ into the internal direct sum $M_0(1)_0 + M_0(1)_1$ constitutes a $\mathbf{Z}_2$-grading of $M_0(1)$. We shall denote $M_0(1)$ graded in this fashion by $M_0(1, 0)$. As discussed in Section 3.3, such a graded associative algebra may be equipped with a bracket $\{\,.\,,\,.\,\}$ defined by bilinear extension of the following rule for arbitrary homogeneous elements A, B of $M_0(1, 0)$:

$$\{A, B\} = A.B - (-1)^{\sigma(A)\sigma(B)} B.A.$$

As in Chapter 3, the function σ is the parity function which maps a homogeneous element to the parity of that element. We write σ_β^α as shorthand for $\sigma(E_\beta^\alpha)$, the parity of the basis element E_β^α of $M_0(1, 0)$. This is defined for all values of α and β. When $M_0(1, 0)$ is equipped with this bracket it becomes an example of a Lie superalgebra, a structure introduced in Chapter 3. This Lie superalgebra is denoted $gl_0(1, 0)$.

Let the quantum stochastic differential process $d\Xi_\beta^\alpha$ be defined for $0 \leq \alpha, \beta \leq 1$ by

$$d\Xi_\beta^\alpha = (-1)^{\sigma_\beta^\alpha \Lambda_1^1} d\Lambda_\beta^\alpha.$$

Hence, if $\sigma_\beta^\alpha = 0$ we have $d\Xi_\beta^\alpha = d\Lambda_\beta^\alpha$ and if $\sigma_\beta^\alpha = 1$ we have $d\Xi_\beta^\alpha = (-1)^{\Lambda_1^1} d\Lambda_\beta^\alpha$. The slightly unusual-looking operator $(-1)^{\Lambda_1^1}$ is derived from the spectral decomposition $\sum_{n=1}^\infty n E_n$ of Λ_1^1 as

$$(-1)^{\Lambda_1^1} = \sum_{n=1}^\infty (-1)^n E_n.$$

At a particular time $t \geq 0$, this process may be considered as the second quantisation of the $L^2(\mathbf{R}_+; \mathbf{C})$ map of pointwise multiplication by the function $-\chi_{[0,t]} + \chi_{(t,\infty)} : \mathbf{R}_+ \to \mathbf{C}$. It is clear from this interpretation that, for a particular time $t \geq 0$, $(-1)^{\Lambda_1^1(t)}$ acts on an arbitrary exponential vector $e(f)$ with f in $L^2(\mathbf{R}_+; \mathbf{C})$ as follows:

$$(-1)^{\Lambda_1^1(t)} e(f) = e(-\chi_{[0,t]} f + \chi_{(t,\infty)} f).$$

Now that we have defined the $d\Xi_\beta^\alpha$ we may define the quantum stochastic processes Ξ_β^α for $0 \leq \alpha, \beta \leq 1$ by means of their constituent operators. Thus, for arbitrary $t \geq 0$ we define $\Xi_\beta^\alpha(t)$ by

$$\Xi_\beta^\alpha(t) = \int_0^t d\Xi_\beta^\alpha(s). \tag{4.6}$$

It was shown in [HP2] that Ξ_1^0 is the Fermionic creation process and that Ξ_0^1 is the Fermionic annihilation process. The relationship between these processes and the gauge and time processes (Ξ_1^1 and Ξ_0^0 respectively) may be summarized by the statement that, for each $t \geq 0$ the map

$$E_\beta^\alpha \mapsto \Xi_\beta^\alpha(t)$$

forms a representation of the Lie superalgebra $gl_0(1,0)$ on the $\mathbf{Z}_2$-graded Hilbert space

$$\Gamma(L^2(\mathbf{R}_+; \mathbf{C})) = \Gamma_{t0} \oplus \Gamma_{t1}.$$

Here Γ_{t0} is the eigenspace of $\Gamma(L^2(\mathbf{R}_+; \mathbf{C}))$ corresponding to the eigenvalue $+1$ of the self-adjoint unitary grading operator $(-1)^{\Lambda_1^1(t)}$. Similarly, Γ_{t1} is the eigenspace corresponding to the eigenvalue -1 of $(-1)^{\Lambda_1^1(t)}$.

It is now clear that for all $t \geq 0$ we have that $\Xi_1^0(t)$ and $\Xi_0^1(t)$ anticommute about $(-1)^{\Lambda_1^1(t)}$ whereas $\Xi_0^0(t)$ and $\Xi_1^1(t)$ commute about $(-1)^{\Lambda_1^1(t)}$. This may be expressed as

$$(-1)^{\Lambda_1^1(t)} \Xi_\beta^\alpha(t) = (-1)^{\sigma_\beta^\alpha} \Xi_\beta^\alpha(t) (-1)^{\Lambda_1^1(t)}.$$

Furthermore, it can be shown that we have the following formal relation for all $\alpha, \beta, \gamma, \delta$ with $0 \leq \alpha, \beta, \gamma, \delta \leq 1$ and all $t \geq 0$:

$$\Xi_\beta^\alpha(t) \Xi_\delta^\gamma(t) - (-1)^{\sigma_\beta^\alpha \sigma_\delta^\gamma} \Xi_\delta^\gamma(t) \Xi_\beta^\alpha(t) = \hat\delta_\delta^\alpha \Xi_\beta^\gamma(t) - (-1)^{\sigma_\beta^\alpha \sigma_\delta^\gamma} \hat\delta_\beta^\gamma \Xi_\delta^\alpha(t). \tag{4.7}$$

If $e(f), e(g)$ are arbitrary exponential vectors with f, g in $L^2(\mathbf{R}_+; \mathbf{C})$ we see that the rigorous form of (4.7) is

$$
\begin{aligned}
&\langle \Xi_\beta^\alpha(t)^\dagger e(f), \Xi_\delta^\gamma(t)e(g)\rangle - (-1)^{\sigma_\beta^\alpha \sigma_\delta^\gamma}\langle \Xi_\delta^\gamma(t)^\dagger e(f), \Xi_\beta^\alpha(t)e(g)\rangle \\
&= \langle e(f), (\hat{\delta}_\delta^\alpha \Xi_\beta^\gamma(t) - (-1)^{\sigma_\beta^\alpha \sigma_\delta^\gamma}\hat{\delta}_\beta^\gamma \Xi_\delta^\alpha(t))e(g)\rangle.
\end{aligned}
\tag{4.8}
$$

It is an N-dimensional generalisation of relation (4.8) that forms the central result of this chapter and, indeed, of this entire work.

4.3 A Partial Generalisation

The Boson–Fermion equivalence of [HP2] reviewed in Sections 2.6 and 4.2 can be easily seen to generalise to N dimensions by $\mathbf{Z}_2$-grading the more general matrix algebra $M_0(N)$ described in Section 2.2 using the decomposition $M_0(N)_0 + M_0(N)_1$ where

$$
M_0(N)_0 = \left\{ \begin{pmatrix} * & 0 & \cdots & 0 \\ 0 & * & \cdots & * \\ \vdots & \vdots & & \vdots \\ 0 & * & \cdots & * \end{pmatrix} \right\}, \quad M_0(N)_1 = \left\{ \begin{pmatrix} 0 & * & \cdots & * \\ * & 0 & \cdots & 0 \\ \vdots & \vdots & & \vdots \\ * & 0 & \cdots & 0 \end{pmatrix} \right\}.
$$

Here the entries marked $*$ take values that are arbitrary complex numbers. This is equivalent to setting

$$
\begin{aligned}
M_0(N)_0 &= \mathrm{span}\left(\{E_j^i : 1 \leq i, j \leq N\} \cup \{E_0^0\}\right), \\
M_0(N)_1 &= \mathrm{span}\left(\{E_0^i, E_i^0 : 1 \leq i \leq N\}\right).
\end{aligned}
$$

It is a simple matter to verify that the inclusions

$$
\begin{aligned}
M_0(N)_0.M_0(N)_0, \quad M_0(N)_1.M_0(N)_1 &\subset M_0(N)_0; \\
M_0(N)_0.M_0(N)_1, \quad M_0(N)_1.M_0(N)_0 &\subset M_0(N)_1
\end{aligned}
$$

hold. Thus the decomposition of $M_0(N)$ into the internal direct sum $M_0(N)_0 + M_0(N)_1$ constitutes a $\mathbf{Z}_2$-grading of $M_0(N)$ and we denote this $\mathbf{Z}_2$-graded associative algebra as $M_0(N, 0)$. When equipped with the superbracket $\{.,.\}$ defined by bilinear extension of the following rule for homogeneous elements A, B of $M_0(N, 0)$:

$$
\{A, B\} = A.B - (-1)^{\sigma(A)\sigma(B)} B.A
$$

the space $M_0(N, 0)$ becomes a Lie superalgebra which we denote $gl_0(N, 0)$.

Let the grading process of Section 4.2 be replaced by $(-1)^{\sum_{j=1}^{N} A_j^j}$ so that we define the differential processes $d\Xi_\beta^\alpha$ for $0 \leq \alpha, \beta \leq N$ by

$$
d\Xi_\beta^\alpha = (-1)^{\sigma_\beta^\alpha} \sum_{j=1}^{N} {}^{A_j^j} d\Lambda_\beta^\alpha
$$

and we define the corresponding processes Ξ_β^α by setting for each $t \geq 0$

$$\Xi_\beta^\alpha(t) = \int_0^t d\Xi_\beta^\alpha(s).$$

This scheme gives a partial N-dimensional generalisation of the Boson–Fermion equivalence result of [HP2]. As in the $M_0(1,0)$ case, the processes Ξ_β^α commute or anticommute about the grading process $(-1)^{\sum_{j=1}^N A_j^j}$ depending on their parity, that is to say that for arbitrary $t \geq 0$,

$$(-1)^{\sum_{j=1}^N A_j^j(t)} \Xi_\beta^\alpha(t) = (-1)^{\sigma_\beta^\alpha} \Xi_\beta^\alpha(t)(-1)^{\sum_{j=1}^N A_j^j(t)}.$$

The obvious generalisations of (4.7) and (4.8) hold. If $t \geq 0$ is arbitrary and $0 \leq \alpha, \beta, \gamma, \delta \leq N$ then formally we have

$$\Xi_\beta^\alpha(t)\Xi_\delta^\gamma(t) - (-1)^{\sigma_\beta^\alpha \sigma_\delta^\gamma} \Xi_\delta^\gamma(t)\Xi_\beta^\alpha(t) = \hat{\delta}_\delta^\alpha \Xi_\beta^\gamma(t) - (-1)^{\sigma_\beta^\alpha \sigma_\delta^\gamma} \hat{\delta}_\beta^\gamma \Xi_\delta^\alpha(t).$$

It is clear what a rigorous version of this relation would be.

4.4 The Construction of $M_0(N,r)$

We will now describe a general means of grading the space $M_0(N)$. Particular examples of this general grading have already been seen in this chapter in the form of $M_0(1,0)$ and $M_0(N,0)$ with their associated Lie superalgebras $gl_0(1,0)$ and $gl_0(N,0)$. We now decompose $M_0(N)$ into the internal direct sum $M_0(N)_0 + M_0(N)_1$ where

$$M_0(N)_0 = \left\{ \begin{array}{c} r \\ r \end{array} \left(\begin{array}{cccccc} * & \cdots & * & 0 & \cdots & 0 \\ \vdots & & \vdots & \vdots & & \vdots \\ * & \cdots & * & 0 & \cdots & 0 \\ 0 & \cdots & 0 & * & \cdots & * \\ \vdots & & \vdots & \vdots & & \vdots \\ 0 & \cdots & 0 & * & \cdots & * \end{array} \right) \right\},$$

$$M_0(N)_1 = \left\{ \begin{array}{c} r \\ r \end{array} \left(\begin{array}{cccccc} 0 & \cdots & 0 & * & \cdots & * \\ \vdots & & \vdots & \vdots & & \vdots \\ 0 & \cdots & 0 & * & \cdots & * \\ * & \cdots & * & 0 & \cdots & 0 \\ \vdots & & \vdots & \vdots & & \vdots \\ * & \cdots & * & 0 & \cdots & 0 \end{array} \right) \right\}. \qquad (4.9)$$

As before, entries marked $*$ may take arbitrary complex values. An equivalent means of defining $M_0(N)_0$ and $M_0(N)_1$ is to set

$$M_0(N)_0 = \text{span}\{E_\beta^\alpha : 0 \le \alpha, \beta \le r \text{ or } r+1 \le \alpha, \beta \le N\} \qquad (4.10)$$

and

$$\begin{aligned} M_0(N)_1 = \text{span}\{E_\beta^\alpha :&(0 \le \alpha \le r \text{ and } r+1 \le \beta \le N) \\ &\text{or } (0 \le \beta \le r \text{ and } r+1 \le \alpha \le N)\}. \end{aligned} \qquad (4.11)$$

In order to establish that this decomposition provides an associative $\mathbf{Z}_2$-graded algebra it is necessary to show that the inclusions

$$\begin{aligned} M_0(N)_0.M_0(N)_0, \ M_0(N)_1.M_0(N)_1 &\subset M_0(N)_0; \\ M_0(N)_0.M_0(N)_1, \ M_0(N)_1.M_0(N)_0 &\subset M_0(N)_1 \end{aligned} \qquad (4.12)$$

hold. This result will be given in the form of a proposition.

Proposition 4.1 The inclusions (4.12) hold.

Proof. While examination of the generic matrices of (4.9) makes it seem fairly clear that the inclusions of (4.12) are true, a more rigorous proof is required. This will be done by considering the four separate cases. By linearity, attention may be restricted to the basis elements E_β^α of $M_0(N)$. If E_β^α and E_δ^γ are the arbitrary basis elements under consideration in each case then their product $E_\beta^\alpha.E_\delta^\gamma$ is equal to $\hat{\delta}_\delta^\alpha E_\beta^\gamma$. Noting that $((0)) \in M_0(N)_0$ and $((0)) \in M_0(N)_1$ we may assume further that $\alpha = \delta \neq 0$.

(i) $M_0(N)_0.M_0(N)_0 \subset M_0(N)_0$
We have that $E_\beta^\alpha.E_\alpha^\gamma = E_\beta^\gamma$ where

$$0 \le \alpha, \beta \le r \text{ or } r+1 \le \alpha, \beta \le N$$

and

$$0 \le \alpha, \gamma \le r \text{ or } r+1 \le \alpha, \gamma \le N.$$

From these inequalities we may deduce that

$$0 \le \alpha, \beta, \gamma \le r \text{ or } r+1 \le \alpha, \beta, \gamma \le N$$

which, in particular means that

$$0 \le \gamma, \beta \le r \text{ or } r+1 \le \gamma, \beta \le N.$$

from which it follows by (4.10) that $E_\beta^\gamma \in M_0(N)_0$.
(ii) $M_0(N)_0.M_0(N)_1 \subset M_0(N)_1$
We have $E_\beta^\alpha.E_\alpha^\gamma = E_\beta^\gamma$ where either

$$\text{(a)} \quad 0 \le \alpha, \beta \le r$$

or

$$\text{(b)} \quad r + 1 \le \alpha, \beta \le N$$

and where exactly one of the following pairs of inequalities holds:

$$\text{(a')} \quad 0 \le \alpha \le r \text{ and } r + 1 \le \gamma \le N$$
$$\text{(b')} \quad 0 \le \gamma \le r \text{ and } r + 1 \le \alpha \le N.$$

It is clear that if (a) holds then (a') holds and so we have $0 \le \beta \le r$ and $r + 1 \le \gamma \le N$ so that $E_\beta^\gamma \in M_0(N)_1$ by (4.11). Otherwise (b) holds and so (b') holds also and therefore $0 \le \gamma \le r$ and $r + 1 \le \beta \le N$ which yields $E_\beta^\gamma \in M_0(N)_1$, again according to (4.11).

(iii) $M_0(N)_1.M_0(N)_0 \subset M_0(N)_1$

This proof is entirely symmetric to that given for (ii) but is included for completeness. We have $E_\beta^\alpha.E_\alpha^\gamma = E_\beta^\gamma$ where precisely one of the following pairs of inequalities holds:

$$\text{(a)} \quad 0 \le \alpha \le r \text{ and } r + 1 \le \beta \le N$$
$$\text{(b)} \quad 0 \le \beta \le r \text{ and } r + 1 \le \alpha \le N$$

and where either

$$\text{(a')} \quad 0 \le \alpha, \gamma \le r$$

or

$$\text{(b')} \quad r + 1 \le \alpha, \gamma \le N.$$

It is clear that if (a') holds then (a) holds and so we have $0 \le \gamma \le r$ and $r+1 \le \beta \le N$ so that $E_\beta^\gamma \in M_0(N)_1$ according to (4.11). Otherwise, (b') holds and so (b) holds also, yielding the inequalities $0 \le \beta \le r$ and $r + 1 \le \gamma \le N$ so that $E_\beta^\gamma \in M_0(N)_1$, again by (4.11).

(iv) $M_0(N)_1.M_0(N)_1 \subset M_0(N)_0$

Again we have $E_\beta^\alpha.E_\alpha^\gamma = E_\beta^\gamma$ where precisely one of

$$\text{(a)} \quad 0 \le \alpha \le r \text{ and } r + 1 \le \beta \le N$$
$$\text{(b)} \quad 0 \le \beta \le r \text{ and } r + 1 \le \alpha \le N$$

holds and where precisely one of

$$\text{(a')} \quad 0 \le \alpha \le r \text{ and } r + 1 \le \gamma \le N$$
$$\text{(b')} \quad 0 \le \gamma \le r \text{ and } r + 1 \le \alpha \le N$$

holds. If (a) is true then it follows that (a') is true and so we have $r + 1 \le \gamma, \beta \le N$. Otherwise (b) is true from which it follows that (b') is true so we have $0 \le \gamma, \beta \le r$. In either case we have $E_\beta^\gamma \in M_0(N)_0$ according to (4.10).

All four inclusions now covered, we may deduce that the proposition holds.∎

Proposition 4.1 allows us to conclude that the decomposition of $M_0(N)$ as the internal direct sum $M_0(N)_0 + M_0(N)_1$ gives an associative $\mathbf{Z}_2$-graded algebra. The parameter r may take any value from 0 to N inclusive although the case $r = N$ yields the trivial grading in which $M_0(N)_0 = M_0(N)$ and $M_0(N)_1 = \{0\}$. With r fixed we denote the appropriate associative superalgebra constructed in the manner discussed in this section by $M_0(N, r)$. A bracket $\{.,.\}$ may be defined on $M_0(N, r)$ by bilinear extension of the following rule for arbitrary homogeneous elements A, B of $M_0(N, r)$:

$$\{A, B\} = A.B - (-1)^{\sigma(A)\sigma(B)} B.A. \tag{4.13}$$

Equipped with this bracket, the space $M_0(N, r)$ becomes a Lie superalgebra which will be denoted $gl_0(N, r)$. In the next section it will be seen how this broad class of Lie superalgebras index $\mathbf{Z}_2$-graded quantum stochastic calculus. Naturally, the function σ in (4.13) denotes the usual parity function that is 0 on elements of the even subspace of $M_0(N, r)$ and 1 on the elements of the odd subspace of $M_0(N, r)$. As usual, σ^α_β denotes the parity $\sigma(E^\alpha_\beta)$ of the basis element E^α_β of $M_0(N, r)$ which, of course, always exists. We conclude this section with a simple but crucial corollary to proposition 4.1.

Corollary 4.2 For a given space $M_0(N, r)$, the following identity holds for arbitrary values of α, β, γ with $0 \le \alpha, \beta, \gamma \le N$:

$$\sigma^\alpha_\beta + \sigma^\gamma_\alpha = \sigma^\gamma_\beta \ (\text{mod } 2).$$

Note that, as $\sigma^\alpha_\beta = \sigma^\beta_\alpha$ for all values of α, β, the equality given in corollary 4.2 may occur in a variety of forms.

4.5 Definition of the Processes of $\mathbf{Z}_2$-Graded Quantum Stochastic Calculus

We now consider the parameters N and r to be fixed once and for all and so fix the Lie superalgebra $gl_0(N, r)$ and the matrix superalgebra $M_0(N, r)$ which retains the . multiplication of (2.2). In order to define the integrator processes of $\mathbf{Z}_2$-graded multidimensional quantum stochastic calculus we must define the generalised grading process which will be denoted G. In the spirit of [HP2] it may be defined by

$$G = (-1)^{\sum_{j=r+1}^{N} A^j_j}. \tag{4.14}$$

Let H denote the $M_0(N, r)$ matrix which is zero everywhere except for on the diagonal where the $(\alpha, \alpha)^{\text{th}}$ entry is $(-1)^{\sigma^\alpha_0}$. Thus we have

$H = \mathrm{diag}(1,\ldots,1,-1,\ldots,-1)$ where there are r '1's and $N-r$ '−1's. Given an arbitrary time $t \geq 0$ and an arbitrary exponential vector $e(f)$ where $f = (f^1,\ldots,f^N) \in L^2(\mathbf{R}_+; \mathbf{C}^N)$, an alternative definition of G equivalent to that of (4.14) can be given by

$$G(t)e(f) = e(\chi_{[0,t]}fH + \chi_{(t,\infty)}f). \tag{4.15}$$

It is clear that, for each $t \geq 0$, $G(t)^2 = Id$. Furthermore, for arbitrary exponential vectors $e(f), e(g)$ with f, g in $L^2(\mathbf{R}_+; \mathbf{C}^N)$ we have

$$\langle G(t)e(f), e(g) \rangle$$
$$= \langle e(\chi_{[0,t]}fH + \chi_{(t,\infty)}f), e(g) \rangle$$
$$= \exp\left(\int_0^t \sum_{i=1}^r f_i(s)g^i(s) - \sum_{i=r+1}^N f_i(s)g^i(s)\, ds + \int_0^t \sum_{i=1}^N f_i(s)g^i(s)\, ds \right)$$
$$= \langle e(f), e(\chi_{[0,t]}gH + \chi_{(t,\infty)}g) \rangle$$
$$= \langle e(f), G(t)e(g) \rangle$$

so we have that for each $t \geq 0$ the operator $G(t)$ is self-adjoint. Thus G is a process consisting of self-adjoint unitary operators.

We are now in a position to define the quantum stochastic differentials $d\Xi_\beta^\alpha$ for arbitrary integers α, β with $0 \leq \alpha, \beta \leq N$ by

$$d\Xi_\beta^\alpha = G^{\sigma_\beta^\alpha}\, d\Lambda_\beta^\alpha. \tag{4.16}$$

We can see that in the case $\sigma_\beta^\alpha = 0$ we have $d\Xi_\beta^\alpha = d\Lambda_\beta^\alpha$. If, however, $\sigma_\beta^\alpha = 1$ then $d\Xi_\beta^\alpha$ is equal to $G\, d\Lambda_\beta^\alpha$. The definition of the differentials given by (4.16) leads to a natural definition of the quantum stochastic processes Ξ_β^α. For arbitrary integers α, β with $0 \leq \alpha, \beta \leq N$ and arbitrary $t \geq 0$ we define the constituent operators of Ξ_β^α by

$$\Xi_\beta^\alpha(t) = \int_0^t d\Xi_\beta^\alpha(s). \tag{4.17}$$

From (4.17) we see that if $\sigma_\beta^\alpha = 0$ then $\Xi_\beta^\alpha = \Lambda_\beta^\alpha$ and if $\sigma_\beta^\alpha = 1$ then $\Xi_\beta^\alpha = \int_0^\cdot G\, d\Lambda_\beta^\alpha$.

In parallel with the ungraded case described in Section 2.2, to each element A of $M_0(N,r)$ there corresponds a process Ξ_A along with its differential $d\Xi_A$. If $A = \lambda_\alpha^\beta E_\beta^\alpha$ for a family of complex numbers $(\lambda_\alpha^\beta, 0 \leq \alpha, \beta \leq N)$ then Ξ_A and $d\Xi_A$ are defined in the obvious way by

$$\Xi_A = \lambda_\alpha^\beta \Xi_\beta^\alpha; \quad d\Xi_A = \lambda_\alpha^\beta d\Xi_\beta^\alpha.$$

We remark that if $A \in M_0(N,r)$ is homogeneous of grade 0 then $\Xi_A = \Lambda_A$ and if A is homogeneous of grade 1 then $\Xi_A = \int_0^\cdot G\, d\Lambda_A$.

4.6 Some Lemmas

Four lemmas will be given in this section all of which will be used in the proof of the theorem given in Section 4.7.

Lemma 4.3 For all $s, t \in \mathbf{R}_+$ we have

$$G(s)G(t) = G(t)G(s).$$

Proof. Assume without loss of generality that $s < t$. For each $s \in \mathbf{R}_+$ the operator $G(s)$ is a self-adjoint unitary so for arbitrary exponential vectors $e(f), e(g)$ with f, g in $L^2(\mathbf{R}_+; \mathbf{C}^N)$ we have that $\langle e(f), G(s)G(t)e(g) \rangle$ is equal to $\langle G(s)e(f), G(t)e(g) \rangle$. By application of (4.15) we have that this inner product is equal to

$$\langle e(\chi_{[0,s]}fH + \chi_{(s,\infty)}f), e(\chi_{[0,t]}gH + \chi_{(t,\infty)}g) \rangle. \tag{4.18}$$

Applying standard Fock space theory as described in Section 2.1 we have that (4.18) is equal to

$$\exp\Big(\int_0^s \sum_{i=1}^N f_i(u)g^i(u)\,du + \int_s^t \sum_{i=1}^r f_i(u)g^i(u)\,du$$
$$- \int_s^t \sum_{i=r+1}^N f_i(u)g^i(u)\,du + \int_t^\infty \sum_{i=1}^N f_i(u)g^i(u)\,du \Big).$$

It is clear that this expression is equal to

$$\langle e(\chi_{[0,t]}fH + \chi_{(t,\infty)}f), e(\chi_{[0,s]}gH + \chi_{(s,\infty)}g) \rangle \tag{4.19}$$

which, from (4.15), is equal to $\langle G(t)e(f), G(s)e(g) \rangle$. The self-adjointness of $G(t)$ yields that this is equal to $\langle e(f), G(t)G(s)e(g) \rangle$. The result now follows from the totality of the exponential vectors in Fock space.■

Lemma 4.4 Given arbitrary $s, t \in \mathbf{R}_+$ with $s \leq t$ and arbitrary integers α, β with $0 \leq \alpha, \beta \leq N$ we have

$$\Xi_\beta^\alpha(s)G(t) = (-1)^{\sigma_\beta^\alpha} G(t)\Xi_\beta^\alpha(s). \tag{4.20}$$

Proof. Note that for each $t \geq 0$, the operator $G(t)$ leaves the exponential domain $\mathcal{E}$ invariant and so the products $\Xi_\beta^\alpha(s)G(t)$ and $G(t)\Xi_\beta^\alpha(s)$ are perfectly rigorous. We consider the cases $\sigma_\beta^\alpha = 0$ and $\sigma_\beta^\alpha = 1$ separately.

If $\sigma_\beta^\alpha = 0$ then for arbitrary exponential vectors $e(f), e(g)$ with f, g in $L^2(\mathbf{R}_+; \mathbf{C}^N)$ we have from (4.15), (4.16) and (4.17) that $\langle e(f), \Xi_\beta^\alpha(s)G(t)e(g) \rangle$ is equal to

$$\langle e(f), \int_0^s d\Lambda_\beta^\alpha(u) e(\chi_{[0,t]} g H + \chi_{(t,\infty)} g) \rangle. \qquad (4.21)$$

By the first fundamental formula given in theorem 2.7, (4.21) is equal to

$$\int_0^s f_\beta(u)(-1)^{\sigma_0^\alpha} g^\alpha(u) \, du \langle e(f), G(t) e(g) \rangle. \qquad (4.22)$$

Note that $(-1)^{\sigma_0^\alpha}$ is 1 if $0 \leq \alpha \leq r$ and -1 if $r < \alpha \leq N$. It is assumed that $\sigma_\beta^\alpha = 0$ and so, according to (4.10), whichever inequality applies to α will also apply to β. It follows from this that $\sigma_0^\alpha = \sigma_\beta^0$. Furthermore, $G(t)$ is a self-adjoint unitary and so (4.22) may be re-written as

$$\int_0^s (-1)^{\sigma_\beta^0} f_\beta(u) g^\alpha(u) \, du \langle G(t) e(f), e(g) \rangle. \qquad (4.23)$$

An application of the first fundamental formula gives us that (4.23) is equal to

$$\langle G(t) e(f), \Xi_\beta^\alpha(s) e(g) \rangle$$

which, by the self-adjointness of $G(t)$, is equal to $\langle e(f), G(t) \Xi_\beta^\alpha(s) e(g) \rangle$. It follows from the totality of the exponential vectors in $\Gamma(L^2(\mathbf{R}_+; \mathbf{C}^N))$ that the lemma holds in the case $\sigma_\beta^\alpha = 0$.

We now attack the alternative case where $\sigma_\beta^\alpha = 1$. As seen in the $\sigma_\beta^\alpha = 0$ case we have by the first fundamental formula that $\langle e(f), \Xi_\beta^\alpha(s) G(t) e(g) \rangle$ is equal to

$$\int_0^s f_\beta(r)(-1)^{\sigma_0^\alpha} g^\alpha(r) \, dr \langle e(f), G(t) e(g) \rangle. \qquad (4.24)$$

If (4.24) is multiplied by $(-1)^{\sigma_\beta^\alpha}$ then corollary 4.2 yields that the equality

$$\langle e(f), (-1)^{\sigma_\beta^\alpha} \Xi_\beta^\alpha(s) G(t) e(g) \rangle = \int_0^s (-1)^{\sigma_\beta^0} f_\beta(u) g^\alpha(u) \, du \langle e(f), G(t) e(g) \rangle. \qquad (4.25)$$

holds. An application of the first fundamental formula along with the self-adjointness of $G(t)$ shows that the right hand side of (4.25) is equal to $\langle G(t) e(f), \Xi_\beta^\alpha(s) e(g) \rangle$ which, again by the self-adjointness of $G(t)$, is equal to $\langle e(f), G(t) \Xi_\beta^\alpha(s) e(g) \rangle$. It follows by the totality of the exponential vectors in $\Gamma(L^2(\mathbf{R}_+; \mathbf{C}^N))$ that the result holds in the case $\sigma_\beta^\alpha = 1$. Both cases now covered, we may conclude that lemma 4.4 holds.∎

We can see from lemma 4.4 that the Ξ_β^α inherit their parity from the corresponding element E_β^α of $M_0(N, r)$. It is clear by linearity that, for an arbitrary homogeneous element A of $M_0(N, r)$, the following generalisation of (4.20) holds:

$$\Xi_A(s) G(t) = (-1)^{\sigma(A)} G(t) \Xi_A$$

for all s, t in $\mathbf{R}_+$ with $s \leq t$.

Lemma 4.5 For arbitrary $s, t \in \mathbf{R}_+$ with $s \leq t$, arbitrary integers $\alpha, \beta, \gamma, \delta$ with $0 \leq \alpha, \beta, \gamma, \delta \leq N$ and arbitrary exponential vectors $e(f), e(g)$ with f, g in $\mathbf{L}^2(\mathbf{R}_+; \mathbf{C}^N)$ we have

$$\langle \Xi_\beta^\alpha(s)^\dagger e(f), (\Xi_\delta^\gamma(t) - \Xi_\delta^\gamma(s))e(g) \rangle =$$
$$(-1)^{\sigma_\beta^\alpha \sigma_\delta^\gamma} \langle (\Xi_\delta^\gamma(t) - \Xi_\delta^\gamma(s))^\dagger e(f), \Xi_\beta^\alpha(s)e(g) \rangle. \tag{4.26}$$

Proof. When $\sigma_\delta^\gamma = 0$, (4.26) follows from the fact that the increments of the process $\Lambda_\delta^\gamma = \Xi_\delta^\gamma$ commute with the past. When $\sigma_\delta^\gamma = 1$ we have from (4.16), (4.17) and the self-adjoint unitarity of the component operators of G that

$$\langle \Xi_\beta^\alpha(s)^\dagger e(f), (\Xi_\delta^\gamma(t) - \Xi_\delta^\gamma(s))e(g) \rangle$$
$$= \langle \Xi_\beta^\alpha(s)^\dagger e(f), \int_s^t G(u) \, d\Lambda_\delta^\gamma(u)e(g) \rangle$$
$$= \langle \Xi_\beta^\alpha(s)^\dagger e(f), G(s) \int_s^t G(s)G(u) \, d\Lambda_\delta^\gamma(u)e(g) \rangle. \tag{4.27}$$

By the self-adjointness of $G(s)$ and lemma 4.4 we may re-write expression (4.27) as

$$(-1)^{\sigma_\beta^\alpha \sigma_\delta^\gamma} \langle \Xi_\beta^\alpha(s)^\dagger G(s)e(f), \int_s^t G(s)G(u) \, d\Lambda_\delta^\gamma(u)e(g) \rangle. \tag{4.28}$$

Recall that we have assumed $\sigma_\delta^\gamma = 1$. As $G(s)$ is a self-adjoint unitary, the $G(s)G(u)$ on the right of the inner product (4.28) commutes with the past up to time s and so (4.28) may, with the help of lemma 4.3, be re-written as

$$(-1)^{\sigma_\beta^\alpha \sigma_\delta^\gamma} \langle \left(\int_s^t G(s)G(u) \, d\Lambda_\delta^\gamma(u) \right)^\dagger G(s)e(f), \Xi_\beta^\alpha(s)e(g) \rangle. \tag{4.29}$$

Moving the $G(s)$ out of the integral, again using lemma 4.3, we may re-write (4.29) as

$$(-1)^{\sigma_\beta^\alpha \sigma_\delta^\gamma} \langle \left(\int_s^t G(u) \, d\Lambda_\delta^\gamma(u) \right)^\dagger e(f), \Xi_\beta^\alpha(s)e(g) \rangle$$

which is clearly equal to

$$(-1)^{\sigma_\beta^\alpha \sigma_\delta^\gamma} \langle (\Xi_\delta^\gamma(t) - \Xi_\delta^\gamma(s))^\dagger e(f), \Xi_\beta^\alpha(s)e(g) \rangle$$

as required.■

We can see from lemma 4.5 that the increments of the process Ξ_β^α super-commute with the past in the weak sense.

Lemma 4.6 For each time $t \geq 0$ and all α, β with $0 \leq \alpha, \beta \leq N$ we have

$$\Xi_\beta^\alpha(t)^\dagger = \Xi_\alpha^\beta(t). \qquad (4.30)$$

Proof. By (4.16) and (4.17)

$$\Xi_\beta^\alpha(t) = \int_0^t G^{\sigma_\beta^\alpha}(s)\, d\Lambda_\beta^\alpha(s)$$

so, from the standard theory of quantum stochastic calculus discussed in Chapter 2, we have

$$\Xi_\beta^\alpha(t)^\dagger = \left(\int_0^t G^{\sigma_\beta^\alpha}(s)\, d\Lambda_\beta^\alpha(s) \right)^\dagger = \int_0^t \left(G^{\sigma_\beta^\alpha}(s) \right)^\dagger d\Lambda_\alpha^\beta(s). \qquad (4.31)$$

We know that both $G(s)$ and Id are self-adjoint and that $\sigma_\beta^\alpha = \sigma_\alpha^\beta$ so we may deduce from (4.31) that $\Xi_\beta^\alpha(t)^\dagger$ is equal to

$$\int_0^t G^{\sigma_\alpha^\beta}(s)\, d\Lambda_\alpha^\beta(s)$$

which is, by (4.16) and (4.17), the process Ξ_α^β as required.∎

In the light of lemma 4.6 it seems reasonable to define an involution on the $\mathbf{Z}_2$-graded quantum stochastic differentials by $(d\Xi_\beta^\alpha)^\dagger = d\Xi_\alpha^\beta$. It is easy to show that this definition satisfies the involution antimultiplicity property under the quantum Ito multiplication. The $\mathbf{Z}_2$-graded version of this multiplication will be given at the end of Section 4.7. We end this section with a corollary which will be needed in Chapter 8.

Corollary 4.7 (To lemma 4.6) For arbitrary integers α, β with $0 \leq \alpha, \beta \leq N$ and arbitrary $t \geq 0$ we have

$$\int (d\Xi_\beta^\alpha)^\dagger = \left(\int_0^t d\Xi_\beta^\alpha \right)^\dagger.$$

Proof. We have that

$$\int_0^t \left(d\Xi_\beta^\alpha \right)^\dagger = \int_0^t d\Xi_\alpha^\beta(s). \qquad (4.32)$$

By (4.17), the right hand term of (4.32) is equal to $\Xi_\alpha^\beta(t)$ which, by lemma 4.6, is equal to $\left(\Xi_\beta^\alpha(t) \right)^\dagger$ which is, by (4.17), the same as $\left(\int_0^t d\Xi_\beta^\alpha \right)^\dagger$ as required.∎

4.7 The Main Theorem

In this section we state the central theorem of $\mathbf{Z}_2$-graded quantum stochastic calculus and give its proof. As in Section 2.2, for a pair of values $s, t \geq 0$ we denote the value $\min\{s, t\}$ by $s \wedge t$.

Theorem 4.8 For a pair of arbitrary exponential vectors $e(f), e(g)$ with $f, g \in \mathbf{L}^2(\mathbf{R}_+; \mathbf{C}^N)$, arbitrary $s, t \in \mathbf{R}_+$ and arbitrary integers $\alpha, \beta, \gamma, \delta$ with $0 \leq \alpha, \beta, \gamma, \delta \leq N$ we have

$$\langle \Xi_\beta^\alpha(s)^\dagger e(f), \Xi_\delta^\gamma(t) e(g) \rangle - (-1)^{\sigma_\beta^\alpha \sigma_\delta^\gamma} \langle \Xi_\delta^\gamma(t)^\dagger e(f), \Xi_\beta^\alpha(s) e(g) \rangle$$
$$= \langle e(f), (\hat{\delta}_\delta^\alpha \Xi_\beta^\gamma(s \wedge t) - (-1)^{\sigma_\beta^\alpha \sigma_\delta^\gamma} \hat{\delta}_\beta^\gamma \Xi_\delta^\alpha(s \wedge t)) e(g) \rangle. \tag{4.33}$$

Before proceeding with the proof we shall make some remarks about this theorem. The theorem gives a rigorous statement of the formal relation

$$\Xi_\beta^\alpha \Xi_\delta^\gamma - (-1)^{\sigma_\beta^\alpha \sigma_\delta^\gamma} \Xi_\delta^\gamma \Xi_\beta^\alpha = \hat{\delta}_\delta^\alpha \Xi_\beta^\gamma - (-1)^{\sigma_\beta^\alpha \sigma_\delta^\gamma} \hat{\delta}_\beta^\gamma \Xi_\delta^\alpha. \tag{4.34}$$

The supercommutator bracket of $gl_0(N, r)$ is denoted by $\{.,.\}$. Let the formal supercommutator bracket of $\mathbf{Z}_2$-graded quantum stochastic processes also be denoted by $\{.,.\}$ so that, for arbitrary processes X, Y of definite grade we have the formal relation $\{X, Y\} = XY - (-1)^{\sigma(X)\sigma(Y)} YX$. Bilinear extension of (4.34) shows that for arbitrary A, B in $gl_0(N, r)$ we have

$$\{\Xi_A, \Xi_B\} = \Xi_{\{A,B\}}.$$

If we consider the process Ξ_A at an arbitrary time $s \geq 0$ and the process Ξ_B at an arbitrary time $t \geq 0$ and denote by $\{.,.\}$ the formal supercommutator of operators defined in the obvious way then the theorem gives

$$\{\Xi_A(s), \Xi_B(t)\} = \Xi_{\{A,B\}}(s \wedge t). \tag{4.35}$$

If A and B are arbitrary elements of $gl_0(N, r)$ which are decomposed into their even and odd parts as $A = A_0 + A_1$ and $B = B_0 + B_1$ respectively then a rigorous and maximal generalisation of the theorem may be given somewhat clumsily as

$$\langle \Xi_{A_0}(s)^\dagger e(f), \Xi_{B_0}(t) e(g) \rangle - \langle \Xi_{B_0}(t)^\dagger e(f), \Xi_{A_0}(s) e(g) \rangle$$
$$+ \langle \Xi_{A_1}(s)^\dagger e(f), \Xi_{B_0}(t) e(g) \rangle - \langle \Xi_{B_0}(t)^\dagger e(f), \Xi_{A_1}(s) e(g) \rangle$$
$$+ \langle \Xi_{A_0}(s)^\dagger e(f), \Xi_{B_1}(t) e(g) \rangle - \langle \Xi_{B_1}(t)^\dagger e(f), \Xi_{A_0}(s) e(g) \rangle \tag{4.36}$$
$$+ \langle \Xi_{A_1}(s)^\dagger e(f), \Xi_{B_1}(t) e(g) \rangle + \langle \Xi_{B_1}(t)^\dagger e(f), \Xi_{A_1}(s) e(g) \rangle$$
$$= \langle e(f), \Xi_{\{A,B\}}(s \wedge t) e(g) \rangle.$$

We remark that the first three pairs of terms on the left of (4.36) correspond to the commutator part of the formal operator bracket $\{\,.\,,.\,\}$ while the fourth pair corresponds to the anticommutator part of this bracket.

These various forms of the theorem all illustrate the fact that the processes Ξ_A yield a time-indexed family of representations of the Lie superalgebra $gl_0(N, r)$. Note that a time-indexed family of representations of the more familiar Lie superalgebra $gl(N, r)$ is embedded in this representation and may be obtained by restricting attention to the pure gauge processes.

Proof. (of theorem 4.8) By lemma 4.5 it suffices to consider the case where $s = t$. By lemma 4.6 we have $\Xi_\beta^\alpha(t)^\dagger = \Xi_\alpha^\beta(t)$ and $\Xi_\delta^\gamma(t)^\dagger = \Xi_\gamma^\delta(t)$. From (4.16) and (4.17) we have for all f, g in $L^2(\mathbf{R}_+; \mathbf{C}^N)$ that

$$\langle \Xi_\alpha^\beta(s)e(f), \Xi_\delta^\gamma(t)e(g)\rangle - (-1)^{\sigma_\beta^\alpha \sigma_\delta^\gamma}\langle \Xi_\gamma^\delta(t)e(f), \Xi_\beta^\alpha(s)e(g)\rangle$$

is equal to

$$\langle \int_0^t G^{\sigma_\alpha^\beta}(s)\, d\Lambda_\alpha^\beta(s)e(f), \int_0^t G^{\sigma_\delta^\gamma}(s)\, d\Lambda_\delta^\gamma(s)e(g)\rangle$$
$$-(-1)^{\sigma_\beta^\alpha \sigma_\delta^\gamma}\langle \int_0^t G^{\sigma_\gamma^\delta}(s)\, d\Lambda_\gamma^\delta(s)e(f), \int_0^t G^{\sigma_\beta^\alpha}(s)\, d\Lambda_\beta^\alpha(s)e(g)\rangle.$$

By the second fundamental formula given in theorem 2.8 this expression is equal to

$$\int_0^t f_\delta(s)g^\gamma(s)\langle \int_0^s G^{\sigma_\alpha^\beta}(u)\, d\Lambda_\alpha^\beta(u)e(f), G^{\sigma_\delta^\gamma}(s)e(g)\rangle\, ds$$
$$+\int_0^t f_\beta(s)g^\alpha(s)\langle G^{\sigma_\alpha^\beta}(s)e(f), \int_0^s G^{\sigma_\delta^\gamma}(u)\, d\Lambda_\delta^\gamma(u)e(g)\rangle\, ds$$
$$+\hat{\delta}_\delta^\alpha \int_0^t f_\beta(s)g^\gamma(s)\langle G^{\sigma_\alpha^\beta}(s)e(f), G^{\sigma_\delta^\gamma}(s)e(g)\rangle\, ds$$
$$-(-1)^{\sigma_\beta^\alpha \sigma_\delta^\gamma}\int_0^t f_\beta(s)g^\alpha(s)\langle \int_0^s G^{\sigma_\gamma^\delta}(u)\, d\Lambda_\gamma^\delta(u)e(f), G^{\sigma_\beta^\alpha}(s)e(g)\rangle\, ds \qquad (4.37)$$
$$-(-1)^{\sigma_\beta^\alpha \sigma_\delta^\gamma}\int_0^t f_\delta(s)g^\gamma(s)\langle G^{\sigma_\gamma^\delta}(s)e(f), \int_0^s G^{\sigma_\beta^\alpha}(u)\, d\Lambda_\beta^\alpha(u)e(g)\rangle\, ds$$
$$-(-1)^{\sigma_\beta^\alpha \sigma_\delta^\gamma}\hat{\delta}_\beta^\gamma \int_0^t f_\delta(s)g^\alpha(s)\langle G^{\sigma_\gamma^\delta}(s)e(f), G^{\sigma_\beta^\alpha}(s)e(g)\rangle\, ds.$$

An application of the first fundamental formula as stated in theorem 2.7 to (4.37) gives the expression

$$\int_0^t f_\delta(s)g^\gamma(s) \int_0^s f_\beta(u)(-1)^{\sigma_\delta^\gamma \sigma_0^\alpha} g^\alpha(u)\langle G^{\sigma_\alpha^\beta}(u)e(f), G^{\sigma_\delta^\gamma}(s)e(g)\rangle \, du \, ds$$

$$+ \int_0^t f_\beta(s)g^\alpha(s) \int_0^s (-1)^{\sigma_\alpha^\beta \sigma_0^0} f_\delta(u)g^\gamma(u)\langle G^{\sigma_\alpha^\beta}(s)e(f), G^{\sigma_\delta^\gamma}(u)e(g)\rangle \, du \, ds$$

$$+ \hat{\delta}_\delta^\alpha \int_0^t f_\beta(s)g^\gamma(s)\langle G^{\sigma_\alpha^\beta}(s)e(f), G^{\sigma_\delta^\gamma}(s)e(g)\rangle \, ds$$

$$- (-1)^{\sigma_\beta^\alpha \sigma_\delta^\gamma} \int_0^t f_\beta g^\alpha \int_0^s f_\delta(u)(-1)^{\sigma_\beta^\alpha \sigma_0^\gamma} g^\gamma(u)\langle G^{\sigma_\gamma^\delta}(u)e(f), G^{\sigma_\beta^\alpha}(s)e(g)\rangle \, du \, ds$$

$$- (-1)^{\sigma_\beta^\alpha \sigma_\delta^\gamma} \int_0^t f_\delta g^\gamma \int_0^s (-1)^{\sigma_\gamma^\delta \sigma_0^\beta} f_\beta(u)g^\alpha(u)\langle G^{\sigma_\gamma^\delta}(s)e(f), G^{\sigma_\beta^\alpha}(u)e(g)\rangle \, du \, ds$$

$$- (-1)^{\sigma_\beta^\alpha \sigma_\delta^\gamma} \hat{\delta}_\beta^\gamma \int_0^t f_\delta(s)g^\alpha(s)\langle G^{\sigma_\gamma^\delta}(s)e(f), G^{\sigma_\beta^\alpha}(s)e(g)\rangle \, ds.$$

$$(4.38)$$

The power of -1 taken by the fifth term of (4.38) is $\sigma_\beta^\alpha \sigma_\delta^\gamma + \sigma_\delta^\gamma \sigma_0^\beta$ which, by corollary 4.2, is equal to $\sigma_\delta^\gamma \sigma_0^\alpha$ modulo 2. It follows from lemma 4.3 and the self-adjoint unitarity of $G^{\sigma_\beta^\alpha}(u)$ and $G^{\sigma_\delta^\gamma}(s)$ that the first and fifth terms of (4.38) cancel. Likewise, the power of -1 taken by the fourth term of (4.38) is $\sigma_\beta^\alpha \sigma_\delta^\gamma + \sigma_\beta^\alpha \sigma_0^\gamma$ which, by corollary 4.2, is equal to $\sigma_\beta^\alpha \sigma_\delta^0$, again modulo 2. It follows from lemma 4.3 and self-adjoint unitarity that the second and fourth terms of (4.38) cancel.

If the third term of (4.38) is to be non-zero we will have by (2.3) that $\alpha = \delta$ so, by corollary 4.2, $\sigma_\beta^\alpha + \sigma_\delta^\gamma = \sigma_\beta^\gamma$. Similarly, if the sixth term of (4.38) is to be non-zero we will have $\gamma = \beta$ so that, by corollary 4.2, $\sigma_\delta^\gamma + \sigma_\beta^\alpha = \sigma_\delta^\alpha$. Thus, recalling that both $G(s)$ and Id are self-adjoint unitary operators, we may re-write (4.38) as

$$\hat{\delta}_\delta^\alpha \int_0^t f_\beta(s)g^\gamma(s)\langle e(f), G^{\sigma_\beta^\gamma}(s)e(g)\rangle \, ds$$

$$- (-1)^{\sigma_\beta^\alpha \sigma_\delta^\gamma} \hat{\delta}_\beta^\gamma \int_0^t f_\delta(s)g^\alpha(s)\langle e(f), G^{\sigma_\delta^\alpha}(s)e(g)\rangle \, ds$$

which, by a single application of the first fundamental formula, can be seen to be equal to

$$\hat{\delta}_\delta^\alpha \langle e(f), \int_0^t G^{\sigma_\beta^\gamma}(s) \, d\Lambda_\beta^\gamma(s)e(g)\rangle - (-1)^{\sigma_\beta^\alpha \sigma_\delta^\gamma} \hat{\delta}_\beta^\gamma \langle e(f), \int_0^t G^{\sigma_\delta^\alpha}(s) \, d\Lambda_\delta^\alpha(s)e(g)\rangle \, ds.$$

$$(4.39)$$

By (4.16), (4.17) and the fact that the inner product is linear on the right, (4.39) is equal to

$$\langle e(f), (\hat{\delta}_\delta^\alpha \Xi_\beta^\gamma(t) - (-1)^{\sigma_\beta^\alpha \sigma_\delta^\gamma} \hat{\delta}_\beta^\gamma \Xi_\delta^\alpha(t))e(g)\rangle$$

and so the theorem holds.∎

Note that this proof, particularly the derivation of (4.39) from (4.37) indicates that Ito multiplication of the differentials $d\Xi^\epsilon_\mu$ should be defined by

$$d\Xi^\alpha_\beta . d\Xi^\gamma_\delta = \hat{\delta}^\alpha_\delta \, d\Xi^\gamma_\beta \qquad (4.40)$$

as would be expected. One might consider the following chain of equalities:

$$d\Xi^\alpha_\beta . d\Xi^\gamma_\delta = G^{\sigma\alpha}_\beta \, d\Lambda^\alpha_\beta . G^{\sigma\gamma}_\delta \, d\Lambda^\gamma_\delta = G^{\sigma\alpha}_\beta {}^{+\sigma\gamma}_\delta \, d\Lambda^\alpha_\beta . d\Lambda^\gamma_\delta$$

$$= \hat{\delta}^\alpha_\delta G^{\sigma\alpha}_\beta {}^{+\sigma\gamma}_\delta \, d\Lambda^\gamma_\beta = \hat{\delta}^\alpha_\delta G^{\sigma\gamma}_\beta \, d\Lambda^\gamma_\beta = \hat{\delta}^\alpha_\delta \, d\Xi^\gamma_\beta . \qquad (4.41)$$

5 The Ungraded Higher Order Ito Product Formula

5.1 The ∗ Product

Let $\mathcal{I}$ be a finite-dimensional complex associative algebra with involution †
and not necessarily containing a unit. We refer to $\mathcal{I}$ as the *Ito algebra*. Let
$\mathcal{T}(\mathcal{I})$ denote the complex vector space of all (ungraded) tensors of finite rank
over $\mathcal{I}$ thus:

$$\mathcal{T}(\mathcal{I}) = \mathbf{C} \oplus \mathcal{I} \oplus (\mathcal{I} \otimes \mathcal{I}) \oplus (\mathcal{I} \otimes \mathcal{I} \otimes \mathcal{I}) \oplus \cdots.$$

An element of $\mathcal{T}(\mathcal{I})$ is a sequence $(a_0, a_1, a_2, \ldots)$ with each a_i an element of
$\mathcal{I} \otimes \cdots \otimes \mathcal{I}$ (i copies of $\mathcal{I}$) and only finitely-many non-zero terms. If we relax
the finiteness condition we obtain the strong sum which we denote $\mathcal{T}_S(\mathcal{I})$. We
define an involution, also denoted †, on $\mathcal{T}(\mathcal{I})$ and $\mathcal{T}_S(\mathcal{I})$ by

$$(a_0, a_1, a_2, \ldots)^\dagger = (a_0{}^\dagger, a_1{}^\dagger, a_2{}^\dagger, \ldots)$$

where the involution † is defined on the n-fold tensor a_n by linear extension
of the following rule for product tensors:

$$(a^1 \otimes \cdots \otimes a^n)^\dagger = a^1{}^\dagger \otimes \cdots \otimes a^n{}^\dagger.$$

For all $n \in \mathbf{N}$ with $n \geq 1$ we define M_n to be the set $\{1, \ldots, n\}$. By convention
we take M_0 to be $\emptyset$. For an arbitrary finite set X we denote the cardinality
of X by $|X|$.

For an arbitrary natural number n take an arbitrary pair of subsets $A = \{i_1 < \cdots < i_{|A|}\}$, $B = \{j_1 < \cdots < j_{|B|}\}$ of M_n such that $A \cup B = M_n$.
From this pair of sets we may define a composition ${}^A\circ{}^B$ which acts between
a tensor in $\mathcal{I} \otimes \cdots \otimes \mathcal{I}$ ($|A|$ copies of $\mathcal{I}$) and a tensor in $\mathcal{I} \otimes \cdots \otimes \mathcal{I}$ ($|B|$
copies of $\mathcal{I}$) yielding a tensor in $\mathcal{I} \otimes \cdots \otimes \mathcal{I}$ (n copies of $\mathcal{I}$). This composition
may be defined by linear extension of the following prescription for product
tensors when both A and B are non-empty:

$$a^1 \otimes \cdots \otimes a^{|A|} {}^A\circ{}^B b^1 \otimes \cdots \otimes b^{|B|} = c^1 \otimes \cdots \otimes c^n \qquad (5.1)$$

where

$$c^m = \begin{cases} a^k & \text{if } i_k = m \in A \cap (M_n \setminus B); \\ b^l & \text{if } j_l = m \in (M_n \setminus A) \cap B; \\ a^k b^l & \text{if } i_k = j_l = m \in A \cap B. \end{cases}$$

The composition $a_0{}^\emptyset \circ^{M_n} b^1 \otimes \cdots \otimes b^n$ is defined to be $a_0\left(b^1 \otimes \cdots \otimes b^n\right)$. Similarly, $a^1 \otimes \cdots \otimes a^n{}^{M_n} \circ^\emptyset b_0$ is defined to be $b_0\left(a^1 \otimes \cdots \otimes a^n\right)$. Naturally $a_0{}^\emptyset \circ^\emptyset b_0$ is defined to be $a_0 b_0$. We remark that, since there are 3^n ways of decomposing the set M_n into an ordered pair (A, B) such that $A \cup B = M_n$, we have 3^n different compositions for each value of n.

The purpose of defining this family of compositions is to allow the definition of the product $*$ in $T_S(\mathcal{I})$. Given arbitrary $a = (a_0, a_1, \ldots), b = (b_0, b_1, \ldots)$ in $T_S(\mathcal{I})$ we define $a * b$ component-wise by

$$(a * b)_n = \sum_{A \cup B = M_n} a_{|A|}{}^A \circ^B b_{|B|} \tag{5.2}$$

where the summation is over the 3^n ways of taking an ordered pair of subsets (A, B) of M_n such that $A \cup B = M_n$. By way of example we give the first three components of $a * b$ in the case where a is of the form $(a_0, a_1, a_2^1 \otimes a_2^2, \ldots)$ and b is of the form $(b_0, b_1, b_2^1 \otimes b_2^2, \ldots)$:

$$(a * b)_0 = a_0 b_0$$
$$(a * b)_1 = a_0 b_1 + a_1 b_1 + a_1 b_0$$
$$(a * b)_2 = a_0 b_2^1 \otimes b_2^2 + a_1 \otimes b_1 + b_1 \otimes a_1 + a_2^1 b_0 \otimes a_2^2 + a_1 b_2^1 \otimes b_2^2$$
$$+ b_2^1 \otimes a_1 b_2^2 + a_2^1 b_1 \otimes a_2^2 + a_2^1 \otimes a_2^2 b_1 + a_2^1 b_2^1 \otimes a_2^2 b_2^2.$$

The nine terms in $(a * b)_2$ correspond to the respective pairs of subsets

$$(\emptyset, \{1, 2\}), (\{1\}, \{2\}), (\{2\}, \{1\}), (\{1, 2\}, \emptyset), (\{1\}, \{1, 2\}),$$
$$(\{2\}, \{1, 2\}), (\{1, 2\}, \{1\}), (\{1, 2\}, \{2\}), (\{1, 2\}, \{1, 2\}).$$

Theorem 5.1 The multiplication $*$ makes $T_S(\mathcal{I})$ into a unital complex associative algebra for which $\dagger$ is an involution and the element $(1, 0, 0, \ldots)$ is the unit element. The space $T(\mathcal{I})$ is a unital $\dagger$-subalgebra of $T_S(\mathcal{I})$.

Proof. It is clear from (5.2) that the multiplication $*$ is bilinear. We prove associativity by considering individual components as follows:

$$((a * b) * c)_n$$

$$= \sum_{D \cup C = M_n} (a * b)_{|D|}{}^D \circ^C c_{|C|}$$

$$= \sum_{D \cup C = M_n} \sum_{A \cup B = M_{|D|}} (a_{|A|}{}^A \circ^B b_{|B|})^D \circ^C c_{|C|}$$

$$= \sum_{A \cup D = M_n} \sum_{B \cup C = M_{|D|}} a_{|A|}{}^A \circ^D (b_{|B|}{}^B \circ^C c_{|C|})_{|D|}$$

$$= \sum_{A \cup D = M_n} a_{|A|}{}^A \circ^D (b * c)_{|D|}$$

$$= (a * (b * c))_n$$

as required.

It is clear from (5.1) and (5.2) that $(a * b)^\dagger = b^\dagger * a^\dagger$ as the involution $\dagger$ in $\mathcal{I}$ must, by definition, satisfy the relation $(cd)^\dagger = d^\dagger c^\dagger$. It is also clear from (5.1) and (5.2) that $(1, 0, 0, \ldots)$ is an identity element. Thus we have that $(\mathcal{T}_S(\mathcal{I}), *)$ is a unital associative $\dagger$-algebra.

The fact that each component $(a*b)_n$ of $(a*b)$ depends only on components of a and b of rank less than or equal to n shows that $\mathcal{T}(\mathcal{I})$ is a unital $\dagger$-subalgebra of $(\mathcal{T}_S(\mathcal{I}), *)$. ∎

5.2 Fundamental Property of *

If we map a list of (ungraded) quantum stochastic differentials $(d\Lambda_1, \ldots, d\Lambda_m)$ to the iterated quantum stochastic integral $\int d\Lambda_1 \ldots d\Lambda_m$ we see that the map is linear in each $d\Lambda_i$. Thus it makes sense to map the ungraded tensor product $d\Lambda_1 \otimes \cdots \otimes d\Lambda_n$ of differentials to the appropriate quantum stochastic integral.

Set the Ito algebra $\mathcal{I}$ of the previous section to be $\{d\Lambda_A : A \in M_0(N)\}$ or else a subalgebra of this algebra. Note that the multiplication in force is that of Ito multiplication of differentials. Let $\mathcal{P}$ be defined as the space of all complex linear combinations of iterated quantum stochastic integrals obtained from elements of $\mathcal{I}$. In view of the previous paragraph we may define a map I^m from $\mathcal{I} \otimes \cdots \otimes \mathcal{I}$ (m copies of $\mathcal{I}$) to $\mathcal{P}$ by linear extension of the following rule for product tensors:

$$I^m : d\Lambda_{A_1} \otimes \cdots \otimes d\Lambda_{A_m} \mapsto \int_{0 < t_1 < \cdots < t_m <\cdot} d\Lambda_{A_1}(t_1) \ldots d\Lambda_{A_m}(t_m).$$

The maps $(I^m, m \geq 0)$ may be amalgamated into a single map I on $\mathcal{T}(\mathcal{I})$ so that we have

$$I((a_0, a_1, \ldots)) = \sum_{m=0}^{\infty} I^m(a_m).$$

Note that the finiteness restriction on $\mathcal{T}(\mathcal{I})$ means that convergence problems do not arise here. By the same token, I cannot be defined on the whole of $\mathcal{T}_S(\mathcal{I})$ because of convergence requirements. For arbitrary $a \in \mathcal{T}(\mathcal{I})$ and arbitrary $t \geq 0$ we write the value of the process $I(a)$ at time t as $I(a)_t$.

The next theorem states that the map I is a 'weak' morphism from the unital complex associative $\dagger$-algebra $(\mathcal{T}(\mathcal{I}), *)$ to $\mathcal{P}$. The unboundedness of the operators under consideration forbids direct multiplication and so we work with adjoints and the inner product. It is for this reason that the morphism must be qualified as 'weak'.

Theorem 5.2 For an arbitrary time $t \geq 0$, arbitrary elements ϕ, ψ of the exponential domain $\mathcal{E}$ and arbitrary $a, b \in \mathcal{T}(\mathcal{I})$ we have

$$\langle I(a)_t^\dagger \phi, I(b)_t \psi \rangle = \langle \phi, I(a * b)_t \psi \rangle. \tag{5.3}$$

Proof. By linearity it suffices to consider the case where a is of the form

$$(0, 0, \ldots, 0, d\Lambda_{\beta_1}^{\alpha_1} \otimes \cdots \otimes d\Lambda_{\beta_l}^{\alpha_l}, 0, \ldots)$$

for some integer $l \geq 0$ and where b is of the form

$$(0, 0, \ldots, 0, d\Lambda_{\delta_1}^{\gamma_1} \otimes \cdots \otimes d\Lambda_{\delta_m}^{\gamma_m}, 0, \ldots)$$

for some integer $m \geq 0$. Linearity also allows us to assume that ϕ, ψ are exponential vectors which we denote $e(f), e(g)$ respectively with f, g being arbitrary elements of $L^2(\mathbf{R}_+; \mathbf{C}^N)$. We denote by $\dot{a}$ the tensor $d\Lambda_{\beta_1}^{\alpha_1} \otimes \cdots \otimes d\Lambda_{\beta_{l-1}}^{\alpha_{l-1}}$ and by $\dot{b}$ the tensor $d\Lambda_{\delta_1}^{\gamma_1} \otimes \cdots \otimes d\Lambda_{\delta_{m-1}}^{\gamma_{m-1}}$. By a slight abuse of notation we declare

$$a = d\Lambda_{\beta_1}^{\alpha_1} \otimes \cdots \otimes d\Lambda_{\beta_l}^{\alpha_l}; \quad b = d\Lambda_{\delta_1}^{\gamma_1} \otimes \cdots \otimes d\Lambda_{\delta_m}^{\gamma_m}.$$

It will sometimes be convenient to write $a = a^1 \otimes \cdots \otimes a^l$, $b = b^1 \otimes \cdots \otimes b^m$.
We now state and prove a lemma.

Lemma 5.3 Using our current notation we have

$$a * b = (a * \dot{b}) \otimes b^m + (\dot{a} * b) \otimes a^l + (\dot{a} * \dot{b}) \otimes a^l b^m.$$

Proof. (Of lemma) Consider an arbitrary component $(a * b)_k$ of $a * b$ where $k > 0$. We have

$$(a * b)_k = \sum_{A \cup B = M_k} a_{|A|}{}^A \circ^B b_{|B|} = \sum_{\substack{A \cup B = M_k \\ |A|=l \; |B|=m}} a^A \circ^B b$$

$$= \sum_{\substack{A \cup D = M_{k-1} \\ |A|=l \; |D|=m-1}} (a^A \circ^D \dot{b}) \otimes b^m + \sum_{\substack{C \cup B = M_{k-1} \\ |C|=l-1 \; |B|=m}} (\dot{a}^C \circ^B b) \otimes a^l$$

$$+ \sum_{\substack{C \cup D = M_{k-1} \\ |C|=l-1 \; |D|=m-1}} (\dot{a}^C \circ^D \dot{b}) \otimes a^l b^m$$

$$= ((a * \dot{b}) \otimes b^m + (\dot{a} * b) \otimes a^l + (\dot{a} * \dot{b}) \otimes a^l b^m)_k$$

and so the lemma is proved.

We now return to the proof of the theorem. If l or m is equal to 0 then (5.3) clearly holds. If $l = m = 1$ then we have

$$\langle I(a)^\dagger_t e(f), I(b)_t e(g) \rangle = \langle \int_0^t d\Lambda^{\beta_1}_{\alpha_1}(s) e(f), \int_0^t d\Lambda^{\gamma_1}_{\delta_1}(s) e(g) \rangle. \tag{5.4}$$

By the second fundamental formula stated in theorem 2.8, the right-hand side of (5.4) is equal to

$$\int_0^t f_{\delta_1}(s) g^{\gamma_1}(s) \langle \int_0^s d\Lambda^{\beta_1}_{\alpha_1}(u) e(f), e(g) \rangle \, ds$$

$$+ \int_0^t f_{\beta_1}(s) g^{\alpha_1}(s) \langle e(f), \int_0^s d\Lambda^{\gamma_1}_{\delta_1}(u) e(g) \rangle \, ds \tag{5.5}$$

$$+ \hat{\delta}^{\alpha_1}_{\delta_1} \int_0^t f_{\beta_1}(s) g^{\gamma_1}(s) \langle e(f), e(g) \rangle \, ds.$$

Taking the adjoint of $\int_0^s d\Lambda^{\beta_1}_{\alpha_1}(u)$ in the first term of (5.5) in order to move it onto the right of the inner product and then applying the first fundamental formula given by theorem 2.7 yields that (5.5) is equal to

$$\langle e(f), \int_0^t \int_0^s d\Lambda^{\alpha_1}_{\beta_1}(u) \, d\Lambda^{\gamma_1}_{\delta_1}(s) e(g) \rangle$$

$$+ \langle e(f), \int_0^t \int_0^s d\Lambda^{\gamma_1}_{\delta_1}(u) \, d\Lambda^{\alpha_1}_{\beta_1}(s) e(g) \rangle \tag{5.6}$$

$$+ \hat{\delta}^{\alpha_1}_{\delta_1} \langle e(f), \int_0^t d\Lambda^{\gamma_1}_{\beta_1}(s) e(g) \rangle.$$

A simple switch to the notation already introduced enables us to re-write (5.6) as

$$\langle e(f), (I(a \otimes b)_t + I(b \otimes a)_t + I(ab)_t) e(g) \rangle. \tag{5.7}$$

It is easily seen from the definition of the * product that (5.7) is equal to

$$\langle e(f), I(a * b)_t e(g)\rangle.$$

We now have that the result holds for all l, m such that $l+m \leq 2$. We proceed by induction. Assume that (5.3) holds for all l, m such that $l + m < k$ for some integer k. Now suppose a, b are such that $l + m = k$. We have

$$\langle I(a)_t^\dagger e(f), I(b)_t e(g)\rangle = \langle \int_0^t I(\dot{a})_s^\dagger \, d\Lambda_{\alpha_l}^{\beta_l}(s)e(f), \int_0^t I(\dot{b})_s \, d\Lambda_{\delta_m}^{\gamma_m}(s)e(g)\rangle. \tag{5.8}$$

By the second fundamental formula, (5.8) is equal to

$$\int_0^t f_{\delta_m}(s)g^{\gamma_m}(s)\langle I(a)_s^\dagger e(f), I(\dot{b})_s e(g)\rangle \, ds$$

$$+ \int_0^t f_{\beta_l}(s)g^{\alpha_l}(s)\langle I(\dot{a})_s^\dagger e(f), I(b)_s e(g)\rangle \, ds \tag{5.9}$$

$$+ \hat{\delta}_{\delta_m}^{\alpha_l} \int_0^t f_{\beta_l}(s)g^{\gamma_m}(s)\langle I(\dot{a})_s^\dagger e(f), I(\dot{b})_s e(g)\rangle \, ds.$$

The inductive hypothesis applies to each of the three inner products so we may re-write (5.9) as

$$\int_0^t f_{\delta_m}(s)g^{\gamma_m}(s)\langle e(f), I(a * \dot{b})_s e(g)\rangle \, ds$$

$$+ \int_0^t f_{\beta_l}(s)g^{\alpha_l}(s)\langle e(f), I(\dot{a} * b)_s e(g)\rangle \, ds \tag{5.10}$$

$$+ \hat{\delta}_{\delta_m}^{\alpha_l} \int_0^t f_{\beta_l}(s)g^{\gamma_m}(s)\langle e(f), I(\dot{a} * \dot{b})_s e(g)\rangle \, ds.$$

Applying the first fundamental formula to (5.10) gives

$$\langle e(f), \int_0^t I(a * \dot{b})_s \, d\Lambda_{\delta_m}^{\gamma_m}(s)e(g)\rangle$$

$$+ \langle e(f), \int_0^t I(\dot{a} * b)_s \, d\Lambda_{\beta_l}^{\alpha_l}(s)e(g)\rangle \tag{5.11}$$

$$+ \langle e(f), \hat{\delta}_{\delta_m}^{\alpha_l} \int_0^t I(\dot{a} * \dot{b})_s \, d\Lambda_{\beta_l}^{\gamma_m}(s)e(g)\rangle.$$

Applying the rule $d\Lambda_{\beta_l}^{\alpha_l}.\Lambda_{\delta_m}^{\gamma_m} = \hat{\delta}_{\delta_m}^{\alpha_l} d\Lambda_{\beta_l}^{\gamma_m}$ to the third term of (5.11) and switching to the more algebraic notation gives us that (5.11) is equal to

$$\langle e(f), I((a * \dot{b}) \otimes b^m)_t e(g)\rangle + \langle e(f), I((\dot{a} * b) \otimes a^l)_t e(g)\rangle$$
$$+ \langle e(f), I((\dot{a} * \dot{b}) \otimes a^l b^m)_t e(g)\rangle. \tag{5.12}$$

Linearity of the inner product on the right and of I gives us that (5.12) is equal to

$$\langle e(f), I((a * \dot{b}) \otimes b^m + (\dot{a} * b) \otimes a^l + (\dot{a} * \dot{b}) \otimes a^l b^m)_t e(g) \rangle$$

which, by lemma 5.3, is equal to

$$\langle e(f), I(a * b)_t e(g) \rangle$$

as required.∎

It can be shown that the map I is injective; the $\mathbf{Z}_2$-graded version of this result is proved in Section 7.7. The space $\mathcal{P}$ consists of the processes $\{I(a): a \in \mathcal{T}(\mathcal{I})\}$ and so we may use ∗ to form a rigorous notion of a product in $\mathcal{P}$. For $A = I(a) \in \mathcal{P}, B = I(b) \in \mathcal{P}$ we define a product $A \odot B$ in $\mathcal{P}$ by

$$A \odot B = I(a * b).$$

It is clear that $A \odot B$ is the element of $\mathcal{P}$ corresponding to the 'weak' product of A and B as determined by means of the adjoint and the inner product. This theory will be described in full detail for the $\mathbf{Z}_2$-graded case in Chapter 7.

6 The Ito Superalgebra

6.1 Preliminary Definitions

Let $\mathcal{I}$ be an arbitrary complex associative superalgebra of finite dimension, not necessarily possessing a unit and possessing a superalgebra involution †️ which, as seen in Section 3.2, must be grade preserving. Thus †️ is a conjugate-linear map from $\mathcal{I}$ to $\mathcal{I}$ such that $(ab)^\dagger = b^\dagger a^\dagger$ and $\sigma(a^\dagger) = \sigma(a)$ where, of course, σ denotes the parity map. We call $\mathcal{I}$ the *Ito superalgebra*. Let $T(\mathcal{I})$ denote the complex vector space of all tensors formed from $\mathcal{I}$:

$$T(\mathcal{I}) = \mathbf{C} \oplus \mathcal{I} \oplus (\mathcal{I} \otimes \mathcal{I}) \oplus (\mathcal{I} \otimes \mathcal{I} \otimes \mathcal{I}) \oplus \cdots.$$

An element of $T(\mathcal{I})$ is a sequence $(a_0, a_1, a_2, \ldots)$ of tensors with each $a_j \in \mathcal{I} \otimes \cdots \otimes \mathcal{I}$ (j copies of $\mathcal{I}$). Such a sequence will only contain a finite number of non-zero terms. If this finiteness restriction is relaxed then we have the strong sum which we denote $T_S(\mathcal{I})$. Sometimes we will denote by $\otimes^i \mathcal{I}$ the tensor product $\mathcal{I} \otimes \cdots \otimes \mathcal{I}$ (i copies of $\mathcal{I}$).

The standard multiplication in both $T(\mathcal{I})$ and $T_S(\mathcal{I})$ is concatenation. This is defined by linear extension of the following rule for a pair of product tensors:

$$(a^1 \otimes \cdots \otimes a^m)(b^1 \otimes \cdots \otimes b^n) = a^1 \otimes \cdots \otimes a^m \otimes b^1 \otimes \cdots \otimes b^n.$$

This product will only be used in Section 6.5. The main aim of this chapter is to define an entirely different product on $T_S(\mathcal{I})$ which will be put to use in Chapters 7 and 8.

The space $T_S(\mathcal{I})$ has a natural $\mathbf{Z}_2$-grading. The parity assigned to a product tensor $a^1 \otimes \cdots \otimes a^n$ with entries of definite parity is [S] $\sigma(a^1) + \cdots + \sigma(a^n)$. Thus we have a $\mathbf{Z}_2$-grading for each of the component spaces $\mathcal{I} \otimes \cdots \otimes \mathcal{I}$. As discussed in [S], the direct sum $a_1 \oplus \cdots \oplus a_n$ of homogeneous elements is only of definite parity if all of the a_i are of the same parity in which case $a_1 \oplus \cdots \oplus a_n$ takes this parity. Thus an element $a = (a_0, a_1, \ldots)$ of $T_S(\mathcal{I})$ is of definite parity if and only if all the a_i are of the same parity. If this is the case then a takes the common parity of all the a_i. In particular, a may only be odd if $a_0 = 0$ as $\mathbf{C}$ has odd space $\{0\}$ and even space $\mathbf{C}$.

The involution †️ on $\mathcal{I}$ must be extended to the whole of $T_S(\mathcal{I})$. This involution will also be denoted †️. It is clear that this involution should be distributive over a sequence $(a_0, a_1, a_2, \ldots)$ in $\mathcal{T}_S(\mathcal{I})$, that is to say $\dagger$ should have the property

$$(a_0, a_1, a_2, \ldots)^\dagger = (a_0{}^\dagger, a_1{}^\dagger, a_2{}^\dagger, \ldots).$$

It remains to establish the involution on each of the spaces $\mathcal{I} \otimes \cdots \otimes \mathcal{I}$. The multiplication with respect to which we must define this involution is the Chevalley multiplication of tensors described in Chapter 3 for the spaces $\mathcal{I} \otimes \cdots \otimes \mathcal{I}$ where there are at least two copies of $\mathcal{I}$. For $\otimes^1 \mathcal{I} = \mathcal{I}$ the involution should clearly be the one with which $\mathcal{I}$ is already equipped. For $\otimes^0 \mathcal{I} = \mathbf{C}$ the involution must be defined with respect to the standard multiplication of complex numbers. It is well-known that complex conjugation is the only possible such involution. For higher products of the form $\otimes^i \mathcal{I}$ with $i > 1$ it suffices by linearity to define an involution on n-fold product tensors with entries of definite parity. We require that

$$((a^1 \otimes \cdots \otimes a^n)(b^1 \otimes \cdots \otimes b^n))^\dagger = (b^1 \otimes \cdots \otimes b^n)^\dagger (a^1 \otimes \cdots \otimes a^n)^\dagger. \quad (6.1)$$

As was described in Chapter 3, multiplication of Chevalley tensors introduces a grading factor that is a power of -1. We must, therefore, define the involution $\dagger$ on the product spaces $\mathcal{I} \otimes \cdots \otimes \mathcal{I}$ so that it introduces a power of -1 itself that results in (6.1) being satisfied. We provide the correct involution in the form of a proposition.

Proposition 6.1 If we define $\dagger$ on the n-fold tensor space $\otimes^n \mathcal{I}$ by linear extension of the following rule for an arbitrary product tensor $a^1 \otimes \cdots \otimes a^n$ with entries of definite parity:

$$(a^1 \otimes \cdots \otimes a^n)^\dagger = (-1)^{\sum_{i<j} \sigma(a^i)\sigma(a^j)} a^1{}^\dagger \otimes \cdots \otimes a^n{}^\dagger \quad (6.2)$$

where $\dagger$ as applied to elements of $\mathcal{I}$ refers to the involution defined on $\mathcal{I}$ then $\dagger$ is an involution on $\otimes^n \mathcal{I}$.

Proof. It suffices to show that $\dagger$ possesses the three properties required of an involution on a superalgebra. First we show that antimultiplicity holds by means of direct calculation on arbitrary product tensors $a^1 \otimes \cdots \otimes a^n$ and $b^1 \otimes \cdots \otimes b^n$ with entries of definite parity:

$$((a^1 \otimes \cdots \otimes a^n)(b^1 \otimes \cdots \otimes b^n))^\dagger$$

$$= ((-1)^{\sum_{i>j} \sigma(a^i)\sigma(b^j)} a^1 b^1 \otimes \cdots \otimes a^n b^n)^\dagger$$

$$= (-1)^{\sum_{i>j} \sigma(a^i)\sigma(b^j)} (-1)^{\sum_{i<j} \sigma(a^i b^i)\sigma(a^j b^j)} b^1{}^\dagger a^1{}^\dagger \otimes \cdots \otimes b^n{}^\dagger a^n{}^\dagger$$

$$= (-1)^{\sum_{i>j} \sigma(a^i)\sigma(b^j) + \sum_{i<j} (\sigma(a^i)+\sigma(b^i))(\sigma(a^j)+\sigma(b^j))} b^1{}^\dagger a^1{}^\dagger \otimes \cdots \otimes b^n{}^\dagger a^n{}^\dagger.$$

$$(6.3)$$

By cancellation, (6.3) is equal to

$$(-1)^{\sum_{i<j}\sigma(a^i)\sigma(a^j)+\sum_{i<j}\sigma(a^i)\sigma(b^j)+\sum_{i<j}\sigma(b^i)\sigma(b^j)}b^{1\dagger}a^{1\dagger}\otimes\cdots\otimes b^{n\dagger}a^{n\dagger}.$$
(6.4)

On the other hand, recalling that the $\dagger$ defined on elements of $\mathcal{I}$ is grade preserving,

$$(b^1\otimes\cdots\otimes b^n)^\dagger(a^1\otimes\cdots\otimes a^n)^\dagger$$

$$=(-1)^{\sum_{i<j}\sigma(b^i)\sigma(b^j)}b^{1\dagger}\otimes\cdots\otimes b^{n\dagger}(-1)^{\sum_{i<j}\sigma(a^i)\sigma(a^j)}a^{1\dagger}\otimes\cdots\otimes a^{n\dagger}$$

$$=(-1)^{\sum_{i<j}\sigma(a^i)\sigma(b^j)+\sum_{i<j}\sigma(a^i)\sigma(a^j)+\sum_{i<j}\sigma(a^i)\sigma(b^j)}b^{1\dagger}a^{1\dagger}\otimes\cdots\otimes b^{n\dagger}a^{n\dagger}$$

which is equal to (6.4).

That $a^{\dagger^\dagger}=a$ for any tensor a is clear from (6.20) and the fact that $\dagger$ as defined on $\mathcal{I}$ is grade preserving. It also follows from the fact that $\dagger$ is grade preserving on $\mathcal{I}$ that $\dagger$ on $\mathcal{I}\otimes\cdots\otimes\mathcal{I}$ is grade preserving.

Thus $\dagger$ satisfies all of the conditions for an involution on $\otimes^n\mathcal{I}$ and the proposition holds.∎

We now have an involution $\dagger$ defined on all of $T_S(\mathcal{I})$.

6.2 Definition of the $^A\!\bullet^B$ Products

In this section we introduce a family of products $^A\!\bullet^B$ defined on certain tensors in $T_S(\mathcal{I})$ and indexed by certain subsets of the integers. It is by means of these products that we shall define the product $\star$ in Section 6.4.

Recall that M_n denotes the set of integers $\{1,\ldots,n\}$ and that for a finite set X we denote the cardinality of this set by $|X|$. Let A and B be subsets of M_n such that $A\cup B=M_n$. The product $^A\!\bullet^B$ is bilinear from $\otimes^{|A|}\mathcal{I}\times\otimes^{|B|}\mathcal{I}$ to $\otimes^n\mathcal{I}$.

We give the definition in two parts, the first relating to the case where at least one of A,B is empty, the second covering the remaining cases. If $n=0$, thus forcing $A=B=\emptyset$, then for $a_0,b_0\in\mathbf{C}$ we define $a_0{}^\emptyset\bullet^\emptyset b_0$ to be a_0b_0. If $A=\emptyset$ and $B=M_n$ with $n\geq 1$ we define the product $a_0{}^\emptyset\bullet^{M_n}b_n$ for $a_0\in\mathbf{C}$ and $b_n\in\mathcal{I}\otimes\cdots\otimes\mathcal{I}$ (n copies of $\mathcal{I}$) by linear extension of the following rule for arbitrary product tensors $b^1\otimes\cdots\otimes b^n$:

$$a_0{}^\emptyset\bullet^{M_n}b^1\otimes\cdots\otimes b^n=a_0(b^1\otimes b^2\otimes\cdots\otimes b^n).$$

In exactly the same way, we define the product $^{M_n}\!\bullet^\emptyset$ between $\mathcal{I}\otimes\cdots\otimes\mathcal{I}$ (n copies of $\mathcal{I}$) and $\mathbf{C}$ by linear extension of the rule

$$a^1\otimes\cdots\otimes a^{n\,M_n}\bullet^\emptyset b_0=b_0(a^1\otimes a^2\otimes\cdots\otimes a^n).$$

We now define the product $^A\!\bullet^B$ for A,B both non-empty. Let $A=\{i_1,\ldots,i_{|A|}\}$ where $i_1<\cdots<i_{|A|}$ and $B=\{j_1,\ldots,j_{|B|}\}$ where $j_1<$

$\cdots < j_{|B|}$. We may now define the product $^A\bullet^B$ between $\mathcal{I}\otimes\cdots\otimes\mathcal{I}$ ($|A|$ copies of $\mathcal{I}$) on the left and $\mathcal{I}\otimes\cdots\otimes\mathcal{I}$ ($|B|$ copies of $\mathcal{I}$) on the right by linear extension of the following rule for an arbitrary $|A|$-fold product tensor $a = a^1\otimes\cdots\otimes a^{|A|}$ with entries in $\mathcal{I}$ of definite parity and an arbitrary $|B|$-fold tensor $b = b^1\otimes\cdots\otimes b^{|B|}$ again with entries in $\mathcal{I}$ of definite parity:

$$a^A \bullet^B b = (-1)^{\sum_{j_k < i_l} \sigma(a^l)\sigma(b^k)} c^1 \otimes\cdots\otimes c^n \qquad (6.5)$$

where, for each $m \in M_n$,

$$c^m = \begin{cases} a^l & \text{if } m = i_l \in A \cap (M_n \setminus B); \\ b^k & \text{if } m = j_k \in (M_n \setminus A) \cap B; \\ a^l b^k & \text{if } m = i_l = j_k \in A \cap B. \end{cases} \qquad (6.6)$$

It is easily shown by induction that each of the sets M_n may be decomposed into an ordered pair of subsets (A, B) such that $A \cup B = M_n$ in 3^n distinct ways. Hence, to each M_n there corresponds a family of 3^n products of the form just described.

We now present an example. Suppose $n = 7$, $A = \{1,4,6,7\}$, $B = \{1,2,3,5,6\}$, $a = a^1 \otimes a^2 \otimes a^3 \otimes a^4$ and $b = b^1 \otimes b^2 \otimes b^3 \otimes b^4 \otimes b^5$. Suppose further that a^2, b^2, b^4 are even and a^1, a^3, a^4, b^1, b^3, b^5 are odd. Then we have from (6.5) that

$$a^1 \otimes a^2 \otimes a^3 \otimes a^{4\{1,4,6,7\}} \bullet^{\{1,2,3,5,6\}} b^1 \otimes b^2 \otimes b^3 \otimes b^4 \otimes b^5$$
$$=(-1)^7 a^1 b^1 \otimes b^2 \otimes b^3 \otimes a^2 \otimes b^4 \otimes a^3 b^5 \otimes a^4$$
$$= - a^1 b^1 \otimes b^2 \otimes b^3 \otimes a^2 \otimes b^4 \otimes a^3 b^5 \otimes a^4.$$

The -1 takes the seventh power because it is precisely the terms $\sigma(a^3)\sigma(b^1)$, $\sigma(a^4)\sigma(b^1)$, $\sigma(a^3)\sigma(b^2)$, $\sigma(a^4)\sigma(b^2)$, $\sigma(a^3)\sigma(b^3)$, $\sigma(a^4)\sigma(b^3)$, $\sigma(a^4)\sigma(b^5)$ of the sum in (6.5) that are 1 and not 0.

In the next section we present a useful computational device for the calculation of the $^A\bullet^B$ products.

6.3 A Computational Device

If $a^1\otimes\cdots\otimes a^n$ and $b^1\otimes\cdots\otimes b^n$ with n a natural number are product tensors with entries of definite parity then their product is, as was seen in Chapter 3, defined to be

$$(-1)^{\sum_{i>j} \sigma(a^i)\sigma(b^j)} a^1 b^1 \otimes\cdots\otimes a^n b^n.$$

We will see in this section that it is possible to use this product in order to produce a convenient means of computing the $^A\bullet^B$ products defined in the previous section.

If $A = \{i_1 < \cdots < i_{|A|}\}$ is a subset of M_n and $a = a^1 \otimes \cdots \otimes a^n$ is a product tensor then we define a formalistic object

$$a^{A \subset M_n} = w^1 \otimes \cdots \otimes w^n$$

by

$$w^k = \begin{cases} a^l & \text{if } k = i_l \in A; \\ 1 & \text{if } k \notin A. \end{cases} \tag{6.7}$$

The 1 in (6.7) is not intended to denote an element of $\mathcal{I}$, even if $\mathcal{I}$ has an identity element. The symbol 1 is simply a computational device with no attached semantic meaning. By convention we take $a1 = 1a = a$, $1^\dagger = 1$ and $\sigma(1) = 0$. If $B = \{j_1 < \cdots < j_{|B|}\} \subset M_n$ is so that $A \cup B = M_n$ and b is a product tensor $b^1 \otimes \cdots \otimes b^{|B|}$ then the formalistic product of tensors

$$a^{A \subset M_n} b^{B \subset M_n} \tag{6.8}$$

yields a tensor which, because of the $A \cup B = M_n$ condition, does not contain the symbol 1 and so is meaningful. If we assume that the entries of a and b are of definite parity then we may write (6.8) as

$$(-1)^{\sum_{i>j} \sigma(w^i)\sigma(\tilde{w}^j)} w^1 \tilde{w}^1 \otimes \cdots \otimes w^n \tilde{w}^n. \tag{6.9}$$

Here $w^1 \otimes \cdots \otimes w^n$ is as above and $\tilde{w}^1 \otimes \cdots \otimes \tilde{w}^n$ is the formal tensor corresponding to $b^{B \subset M_n}$. The factor of -1 is obtained by direct application of definition (6.7) and the definition of the multiplication of tensors. Direct application of the definitions shows that each $w^m \tilde{w}^m$ will be

$$a^k \text{ if } m = i_k \in A \text{ and } m \notin B,$$
$$b^l \text{ if } m = j_l \in B \text{ and } m \notin A,$$
$$a^k b^l \text{ if } m = i_k = j_l \in A \cap B.$$

Thus (6.8) and (6.9) correspond to the rigorous definition of $^A \bullet^B$ given by (6.5) and (6.6). The example given in the next paragraph illustrates the computational advantage provided by the 'fictitious 1'.

Take, as in Section 6.2, $n = 7$, $A = \{1, 4, 6, 7\}$, $B = \{1, 2, 3, 5, 6\}$, $a = a^1 \otimes a^2 \otimes a^3 \otimes a^4$ and $b = b^1 \otimes b^2 \otimes b^3 \otimes b^4 \otimes b^5$. We take a^2, b^2, b^4 to be even and a^1, a^3, a^4, b^1, b^3, b^5 to be odd. Then we have

$$a^{\{1,4,6,7\} \subset M_7} = a^1 \otimes 1 \otimes 1 \otimes a^2 \otimes 1 \otimes a^3 \otimes a^4$$
$$b^{\{1,2,3,5,6\} \subset M_7} = b^1 \otimes b^2 \otimes b^3 \otimes 1 \otimes b^4 \otimes b^5 \otimes 1.$$

The product of tensors

$$a^{A \subset M_7} b^{B \subset M_7} = (a^1 \otimes 1 \otimes 1 \otimes a^2 \otimes 1 \otimes a^3 \otimes a^4)(b^1 \otimes b^2 \otimes b^3 \otimes 1 \otimes b^4 \otimes b^5 \otimes 1)$$

is, by direct application of the rule given in Section 3.2, equal to

$$(-1)^S a^1 b^1 \otimes b^2 \otimes b^3 \otimes a^2 \otimes b^4 \otimes a^3 b^5 \otimes a^4. \tag{6.10}$$

Here S is equal to

$$
\begin{aligned}
&\sigma(b^1)(\sigma(a^4) + \sigma(a^3) + \sigma(1) + \sigma(a^2) + \sigma(1) + \sigma(1)) \\
+&\sigma(b^2)(\sigma(a^4) + \sigma(a^3) + \sigma(1) + \sigma(a^2) + \sigma(1)) \\
+&\sigma(b^3)(\sigma(a^4) + \sigma(a^3) + \sigma(1) + \sigma(a^2)) \\
+&\sigma(1)(\sigma(a^4) + \sigma(a^3) + \sigma(1)) \\
+&\sigma(b^4)(\sigma(a^4) + \sigma(a^3)) \\
+&\sigma(b^5)\sigma(a^4).
\end{aligned}
$$

By using the parities already assigned to the entries of a and b and the fact that we have taken $\sigma(1)$ to be 0 we see that S is equal to 7 as follows:

$$
\begin{aligned}
S =& 1(1+1+0+0+0+0) + 1(1+1+0+0+0) + 1(1+1+0+0) \\
&+0(1+1+0) + 0(1+1) + 1(1) = 2+2+2+0+0+1 = 7.
\end{aligned}
$$

A rigorous and useful implementation of the device described in this section may be found in [HPu].

6.4 The $\star$ Product

The $\star$ product is defined on the whole of $T_S(\mathcal{I})$. If $a, b \in T_S(\mathcal{I})$ with $a = (a_0, a_1, \ldots)$ and $b = (b_0, b_1, \ldots)$ then we define the product $a \star b$ component-wise by setting for each $n \geq 0$

$$(a \star b)_n = \sum_{A \cup B} a_{|A|}{}^A \bullet^B b_{|B|}. \tag{6.11}$$

Thus $a \star b$ may be written as

$$\left(a_0 b_0, \sum_{A \cup B = M_1} a_{|A|}{}^A \bullet^B b_{|B|}, \sum_{A \cup B = M_2} a_{|A|}{}^A \bullet^B b_{|B|}, \ldots \right).$$

The sum on the right of (6.11) is over all 3^n ways of decomposing M_n into ordered pairs of subsets (A, B) such that $A \cup B = M_n$. The $\star$ product is described for the case where $\mathcal{I} = \{d\Xi_A : A \in M_0(1,0)\}$ in [HS]. We shall now state and prove a theorem concerning this product.

Theorem 6.2 When equipped with the multiplication $\star$ the space $T_S(\mathcal{I})$ is a unital complex associative superalgebra with involution $\dagger$ and unit element $(1, 0, 0, \ldots)$.

Proof. First we establish associativity. If $a, b, c \in T_S(\mathcal{I})$ then for all $n \geq 0$ we have

$$((a \star b) \star c)_n = \sum_{D \cup C = M_n} (a \star b)_{|D|}{}^D \bullet^C c_{|C|}$$

$$= \sum_{D \cup C = M_n} \sum_{A \cup B = M_{|D|}} (a_{|A|}{}^A \bullet^B b_{|B|})^D \bullet^C c_{|C|}$$

$$= \sum_{A \cup D = M_n} \sum_{B \cup C = M_{|D|}} a_{|A|}{}^A \bullet^D (b_{|B|}{}^B \bullet^C c_{|C|})$$

$$= \sum_{A \cup D = M_n} a_{|A|}{}^A \bullet^D (b \star c)_{|D|}$$

$$= (a \star (b \star c))_n$$

and so ⋆ is associative.

We now show that the relation $(a \star b)^\dagger = b^\dagger \star a^\dagger$ holds. We can see from (6.11) that it suffices to show for arbitrary n, for arbitrary $A, B \subset M_n$ such that $A \cup B = M_n$ and arbitrary product tensors $a = a^1 \otimes \cdots \otimes a^{|A|}$, $b = b^1 \otimes \cdots \otimes b^{|B|}$ with entries of definite parity that

$$(a^A \bullet^B b)^\dagger = b^{\dagger B} \bullet^A a^\dagger. \tag{6.12}$$

From (6.2) and (6.5) we can see that the entries of the product tensors on either side of (6.12) will be the same. It remains to show that the powers of -1 introduced on either side of (6.12) by (6.2) and (6.5) are the same. On the left hand side, (6.5) gives a factor of $(-1)^{\sum_{j_k < i_l} \sigma(a^l)\sigma(b^k)}$. This must be multiplied by the factor introduced by (6.2) which is

$$(-1)^{\sum_{i<j} \sigma(a^i)\sigma(a^j) + \sum_{i<j} \sigma(b^i)\sigma(b^j) + \sum_{i_l \neq j_k} \sigma(a^l)\sigma(b^k)}$$

so that the overall factor on the left of (6.12) is

$$(-1)^{\sum_{i<j} \sigma(a^i)\sigma(a^j) + \sum_{i<j} \sigma(b^i)\sigma(b^j) + \sum_{i_l < j_k} \sigma(a^l)\sigma(b^k)}. \tag{6.13}$$

On the right-hand side of (6.12) we have a factor of $(-1)^{\sum_{i<j} \sigma(b^i)\sigma(b^j)}$ introduced by applying (6.2) to b, a factor of $(-1)^{\sum_{i<j} \sigma(a^i)\sigma(a^j)}$ introduced by applying (6.2) to a and, recalling that $\dagger$ is grade preserving, a factor of $(-1)^{\sum_{i_l < j_k} \sigma(a^l)\sigma(b^k)}$ introduced by (6.5). Thus the overall factor on the right of (6.12) is

$$(-1)^{\sum_{i<j} \sigma(a^i)\sigma(a^j) + \sum_{i<j} \sigma(b^i)\sigma(b^j) + \sum_{i_l < j_k} \sigma(a^l)\sigma(b^k)} \tag{6.14}$$

which is the same as the left-hand factor (6.13).

It is clear from (6.11) that $(1, 0, 0, \ldots)$ is a unit for $(\mathcal{T}_S(\mathcal{I}), \star)$ so it remains to show that $(\mathcal{T}_S(\mathcal{I}), \star)$ is a superalgebra. Let A, B be arbitrary subsets of M_n such that $A \cup B = M_n$. If $a = a^1 \otimes \cdots \otimes a^{|A|}$ and $b = b^1 \otimes \cdots \otimes b^{|B|}$ have entries of definite parity then the the parity $\sigma(a)$ of a is $\sigma(a^1) + \cdots + \sigma(a^{|A|})$ (mod 2) and the parity $\sigma(b)$ of b is $\sigma(b^1) + \cdots + \sigma(b^{|B|})$ (mod 2). It is clear from (6.5) and (6.6) that the parity of the product $a^A \bullet^B b$ is independent of A, B and is given by $\sigma(a^1) + \cdots + \sigma(a^{|A|}) + \sigma(b^1) + \cdots + \sigma(b^{|B|}) = \sigma(a) + \sigma(b)$ (mod 2). By linearity we may conclude that for arbitrary homogeneous $a, b \in T_S(\mathcal{I})$ we have $\sigma(a \star b) = \sigma(a) + \sigma(b)$ (mod 2). This result might be expressed as

$$T_S(\mathcal{I})_0 \star T_S(\mathcal{I})_0, \ T_S(\mathcal{I})_1 \star T_S(\mathcal{I})_1 \subset T_S(\mathcal{I})_0;$$
$$T_S(\mathcal{I})_0 \star T_S(\mathcal{I})_1, \ T_S(\mathcal{I})_1 \star T_S(\mathcal{I})_0 \subset T_S(\mathcal{I})_1$$

which is the condition given in (3.1) for $(T_S(\mathcal{I}), \star)$ to be a superalgebra.

All properties asserted have now been proved and so the theorem holds.∎

Corollary 6.3 The space $T(\mathcal{I})$ is a unital associative superalgebra with involution † under the multiplication $\star$.

Proof. From (6.11) we see that each component $(a \star b)_n$ of $a \star b$ depends only on components of a and b of rank no greater than n so the result follows.∎

6.5 Supersymmetric Tensors

As noted in Section 6.1, the usual product in $T_S(\mathcal{I})$ is concatenation. Denote by $\mathcal{R}$ the two-sided ideal of $T_S(\mathcal{I})$ generated under this product by all elements of the form

$$a \otimes b - (-1)^{\sigma(a)\sigma(b)} b \otimes a \tag{6.15}$$

where a and b are arbitrary homogeneous elements of $\mathcal{I}$. We may now define the *strong supersymmetric superalgebra* $S_S(\mathcal{I})$ of $T_S(\mathcal{I})$ by

$$S_S(\mathcal{I}) = \frac{T_S(\mathcal{I})}{\mathcal{R}}.$$

If we denote the coset of $T_S(\mathcal{I})/\mathcal{R}$ corresponding to an arbitrary element a of $T_S(\mathcal{I})$ by $[a]$ then we may define a map $\imath$ from $S_S(\mathcal{I})$ to $T_S(\mathcal{I})$ by linear extension of the following rule for product elements $a^1 \otimes \cdots \otimes a^n$ of $T_S(\mathcal{I})$ with entries of definite parity:

$$\imath : [a^1 \otimes \cdots \otimes a^n] \mapsto \sum_{\pi \in S_n} (-1)^{\sum_{\substack{j<k \\ \pi(j)>\pi(k)}} \sigma(a^{\pi(j)})\sigma(a^{\pi(k)})} a^{\pi(1)} \otimes \cdots \otimes a^{\pi(n)}.$$

$$\tag{6.16}$$

This embedding is motivated by the nature of the quotient space $T_S(\mathcal{I})/\mathcal{R}$. If τ is the permutation $(i\ i+1)$ where i is an arbitrary integer with $1 \leq i \leq n-1$ then this quotient, heuristically speaking, identifies the product tensor having homogeneous entries $a^1 \otimes \cdots \otimes a^n$ with the product tensor $(-1)^{\sigma(a^i)\sigma(a^{i+1})} a^{\tau(1)} \otimes \cdots \otimes a^{\tau(n)}$. Thus, for an arbitrary permutation $\pi \in S_n$, the factor applied to $a^{\pi(1)} \otimes \cdots \otimes a^{\pi(n)}$ in order for it to lie in the same coset as $a^1 \otimes \cdots \otimes a^n$ depends on the number modulo 2 of exchanges of adjacent pairs of odd elements of $a^1 \otimes \cdots \otimes a^n$ that must take place in order to 'reach' the configuration $a^{\pi(1)} \otimes \cdots \otimes a^{\pi(n)}$. If a pair of not necessarily adjacent entries a^i and a^j are such that a^i lies to the left of a^j in $a^1 \otimes \cdots \otimes a^n$ and a^i lies to the right of a^j in $a^{\pi(1)} \otimes \cdots \otimes a^{\pi(n)}$ then it is clear that, no matter by what sequence of exchanges the change of configuration has been made, an odd number of exchanges of a^i and a^j must have occurred and so a factor of $(-1)^{\sigma(a^i)\sigma(a^j)}$ must be introduced. If, on the other hand, a^i lies to the left of a^j in both $a^1 \otimes \cdots \otimes a^n$ and $a^{\pi(1)} \otimes \cdots \otimes a^{\pi(n)}$ then, no matter by what sequence of exchanges the change of configuration has been made, an even number of exchanges of a^i and a^j must have occurred and so no factor need be introduced. In the product tensor $a^1 \otimes \cdots \otimes a^n$, the elements a^i, a^j occupy the ith and j^{th} positions respectively. In the product tensor $a^{\pi(1)} \otimes \cdots \otimes a^{\pi(n)}$ the elements a^i, a^j occupy the $\pi^{-1}(i)^{\text{th}}$ and $\pi^{-1}(j)^{\text{th}}$ positions respectively. Thus, a factor of $(-1)^{\sigma(a^i)\sigma(a^j)}$ must be introduced in the case $i < j$ and $\pi^{-1}(i) > \pi^{-1}(j)$. We may therefore conclude that for an arbitrary permutation $\pi \in S_n$, the product tensor $a^1 \otimes \cdots \otimes a^n$ lies in the same coset of $\mathcal{S}_S(\mathcal{I})$ as

$$(-1)^{\sum_{\substack{i<j \\ \pi^{-1}(i)>\pi^{-1}(j)}} \sigma(a^i)\sigma(a^j)} a^{\pi(1)} \otimes \cdots \otimes a^{\pi(n)}. \qquad (6.17)$$

This may be re-written as

$$(-1)^{\sum_{\substack{\pi(i)<\pi(j) \\ i>j}} \sigma(a^{\pi(i)})\sigma(a^{\pi(j)})} a^{\pi(1)} \otimes \cdots \otimes a^{\pi(n)}$$

which, by simple exchange of the labels i and j, is equal to

$$(-1)^{\sum_{\substack{i<j \\ \pi(i)>\pi(j)}} \sigma(a^{\pi(i)})\sigma(a^{\pi(j)})} a^{\pi(1)} \otimes \cdots \otimes a^{\pi(n)}. \qquad (6.18)$$

Thus the motivation for (6.16) can be seen.

We claim that $\imath$ is an embedding of $\mathcal{S}_S(\mathcal{I})$ into $\mathcal{T}_S(\mathcal{I})$. Take an arbitrary product tensor $a^1 \otimes \cdots \otimes a^n$ with entries of definite parity and an arbitrary integer i with $1 \leq i \leq n-1$. Again we write τ for the permutation $(i \; i+1)$ in S_n. The embedding property will be proved if we can show that

$$\imath([a^1 \otimes \cdots \otimes a^n]) = \imath([(-1)^{\sigma(a^i)\sigma(a^{i+1})} a^{\tau(1)} \otimes \cdots \otimes a^{\tau(n)}]). \qquad (6.19)$$

Bearing in mind the discussion of the previous paragraph, we note that on the right hand side of (6.19) an element a^i will occupy the $\tau(i)^{\text{th}}$ position. After application of an arbitrary permutation $\pi \in S_n$, the position of a^i on the right hand side of (6.19) will be $\tau\pi^{-1}(i)$. Thus $\imath([(-1)^{\sigma(a^i)\sigma(a^{i+1})} a^{\tau(1)} \otimes \cdots \otimes a^{\tau(n)}])$ may be written as

$$(-1)^{\sigma(a^i)\sigma(a^{i+1})} \sum_{\pi \in S_n} (-1)^{\displaystyle \sum_{\substack{\tau(j)<\tau(k) \\ \tau\pi^{-1}(j)>\tau\pi^{-1}(k)}} \sigma(a^j)\sigma(a^k)} \, a^{\pi\tau(1)} \otimes \cdots \otimes a^{\pi\tau(n)}.$$

(6.20)

Our aim is to manipulate (6.20) so that the $\tau(j) < \tau(k)$ clause becomes $j < k$. This will enable us to show that (6.19) holds. We do this by considering all possible values of j and k. If both j and k are not elements of the set $\{i, i+1\}$ then it is clear that the condition $\tau(j) < \tau(k)$ is equivalent to $j < k$. There are four separate cases to consider where precisely one of j, k is an element of $\{i, i+1\}$.

(i) $j = i$, $k \notin \{i, i+1\}$: The condition $\tau(j) < \tau(k)$ becomes $i+1 < k$. We have by assumption that $k \neq i+1$ so the inequality is equivalent to $i < k$, which is the same as $j < k$ as required.

(ii) $j = i+1$, $k \notin \{i, i+1\}$: The condition $\tau(j) < \tau(k)$ becomes $i < k$. We have by assumption that $k \neq i+1$ so the inequality is equivalent to $i+1 < k$, which is the same as $j < k$ as required.

(iii) $j \notin \{i, i+1\}$, $k = i$: The condition $\tau(j) < \tau(k)$ becomes $j < i+1$. We have by assumption that $j \neq i$ so the inequality is equivalent to $j < i$, which is the same as $j < k$ as required.

(iv) $j \notin \{i, i+1\}$, $k = i+1$: The condition $\tau(j) < \tau(k)$ becomes $j < i$. We have by assumption that $j \neq i$ so the inequality is equivalent to $j < i+1$, which is the same as $j < k$ as required.

It remains to deal with the case $j = i+1$, $k = i$, which satisfies the condition $\tau(j) < \tau(k)$. This must be done by considering two possibilities relating to the action of a given permutation π in S_n.

Suppose π is such that $\tau\pi^{-1}(i+1) > \tau\pi^{-1}(i)$. Then a term $\sigma(a^i)\sigma(a^{i+1})$ appears in the sum under consideration. If the summation clause $\tau(j) < \tau(k)$ is changed to $j < k$ the case $j = i+1$, $k = i$ will be barred and instead the case $j = i$, $k = i+1$ will be admitted. The second clause will not be satisfied by this and so the term $\sigma(a^i)\sigma(a^{i+1})$ will be omitted.

Suppose, on the other hand, that π is such that $\tau\pi^{-1}(i+1) < \tau\pi^{-1}(i)$. Then the term $\sigma(a^i)\sigma(a^{i+1})$ does not appear in the sum under consideration. If the summation clause $\tau(j) < \tau(k)$ is changed to $j < k$ the case $j = i$, $k = i+1$ will be admitted and the second clause will be satisfied by this pair of values. This means that the term $\sigma(a^i)\sigma(a^{i+1})$ will be admitted.

Having now covered all possibilities of the values of j and k and determined that a 'correction' of $\sigma(a^i)\sigma(a^{i+1})$ must be introduced for each permutation π in S_n, we can see that the sum

$$\sum_{\substack{\tau(j)<\tau(k) \\ \tau\pi^{-1}(j)>\tau\pi^{-1}(k)}} \sigma(a^j)\sigma(a^k)$$

will be equal to

$$\sigma(a^i)\sigma(a^{i+1}) + \sum_{\substack{j<k \\ \tau\pi^{-1}(j)>\tau\pi^{-1}(k)}} \sigma(a^j)\sigma(a^k).$$

This result enables us to rewrite (6.20), with Y defined to be $(-1)^{\sigma(a^i)\sigma(a^{i+1})}$ for typographical convenience, as

$$Y \sum_{\pi \in S_n} (-1)^{\sigma(a^i)\sigma(a^{i+1}) + \sum_{\substack{j<k \\ \tau\pi^{-1}(j)>\tau\pi^{-1}(k)}} \sigma(a^j)\sigma(a^k)} a^{\pi\tau(1)} \otimes \cdots \otimes a^{\pi\tau(n)}.$$

(6.21)

Cancelling the $(-1)^{\sigma(a^i)\sigma(a^{i+1})}$ and substituting $\pi\tau(j)$ for j and $\pi\tau(k)$ for k we see that (6.21) is equal to

$$\sum_{\pi \in S_n} (-1)^{\sum_{\substack{\pi\tau(j)<\pi\tau(k) \\ j>k}} \sigma(a^{\pi\tau(j)})\sigma(a^{\pi\tau(k)})} a^{\pi\tau(1)} \otimes \cdots \otimes a^{\pi\tau(n)}. \qquad (6.22)$$

The fact that S_n is a group enables us to re-write (6.22) as

$$\sum_{\pi \in S_n} (-1)^{\sum_{\substack{\pi(j)<\pi(k) \\ j>k}} \sigma(a^{\pi(j)})\sigma(a^{\pi(k)})} a^{\pi(1)} \otimes \cdots \otimes a^{\pi(n)}$$

which, by a simple exchange of the labels j and k, is equal to

$$\sum_{\pi \in S_n} (-1)^{\sum_{\substack{j<k \\ \pi(j)>\pi(k)}} \sigma(a^{\pi(j)})\sigma(a^{\pi(k)})} a^{\pi(1)} \otimes \cdots \otimes a^{\pi(n)}. \qquad (6.23)$$

By direct application of (6.16) we have that (6.23) is equal to $\imath([a^1 \otimes \cdots \otimes a^n])$ and so we have that $\imath$ is indeed an embedding of $\mathcal{S}_S(\mathcal{I})$ into $\mathcal{T}_S(\mathcal{I})$. Thus $\mathcal{S}_S(\mathcal{I})$ may be identified with $\imath(\mathcal{S}_S(\mathcal{I})) \subset \mathcal{T}_S(\mathcal{I})$.

If $\mathcal{Q}$ is defined to be the two-sided ideal (under the concatenation product) of $\mathcal{T}(\mathcal{I})$ generated by elements of the form $a \otimes b - (-1)^{\sigma(a)\sigma(b)} b \otimes a$ where a, b are arbitrary homogeneous elements of $\mathcal{I}$ then the space $\mathcal{T}(\mathcal{I})/\mathcal{Q}$ is called the *supersymmetric superalgebra* of $\mathcal{T}(\mathcal{I})$ and is denoted $\mathcal{S}(\mathcal{I})$. This space may be identified with its image under the appropriate restriction of $\imath$ and thought of as $\mathcal{S}_S(\mathcal{I}) \cap \mathcal{T}(\mathcal{I})$.

Theorem 6.4 The strong supersymmetric superalgebra $\mathcal{S}_S(\mathcal{I})$ defined above is a unital associative sub-†-superalgebra of $\mathcal{T}_S(\mathcal{I})$ under the multiplication $\star$.

Proof. That $\mathcal{S}_S(\mathcal{I})$ is closed under † is clear. We have seen that $(\mathcal{T}_S(\mathcal{I}), \star)$ is associative and that it has unit $(1, 0, 0, \ldots)$. This unit is clearly an element of $\mathcal{S}_S(\mathcal{I})$. It remains to show that space $\mathcal{S}_S(\mathcal{I})$ is closed under $\star$.

We begin by considering the product

$$\imath([a^1 \otimes \cdots \otimes a^m]) \star (0, b, 0, 0, \ldots)$$

where b and each a^i are arbitrary homogeneous elements of $\mathcal{I}$ and m is an integer greater than 1. It is clear that $(0, b, 0, 0, \ldots) = \imath([(0, b, 0, 0, \ldots)])$ is a supersymmetric tensor. By (6.11) we have for each $n \geq 0$

$$(\imath([a^1 \otimes \cdots \otimes a^m]) \star (0, b, 0, 0, \ldots))_n =$$

$$\sum_{A \cup B = M_n} \imath([a^1 \otimes \cdots \otimes a^m])_{|A|}{}^A \bullet^B (0, b, 0, 0, \ldots)_{|B|}.$$

As only the first component of $(0, b, 0, 0, \ldots)$ is possibly non-zero we may restrict our attention to those decompositions (A, B) of M_n for which $|B| = 1$, i.e., those for which B is a singleton set. Furthermore, $\imath([a^1 \otimes \cdots \otimes a^m])$ is an m-fold tensor and so we must have $|A| = m$. This restriction leads to the establishment of two distinct sums. The first sum corresponds to the family of decompositions of M_m in which $A = M_m$ and $B = \{i\}$ with $i = 1, \ldots, m$. The second sum corresponds to the family of decompositions of M_{m+1} in which $A = M_{m+1} \setminus \{i\}$ and $B = \{i\}$ with $i = 1, \ldots, m+1$.

The first sum may be written as

$$\sum_{i=1}^{m} \imath([a^1 \otimes \cdots \otimes a^m])^{M_m} \bullet^{\{i\}} b. \tag{6.24}$$

As a matter of convenience we define Y_π for each $\pi \in S_n$ by

$$Y_\pi = (-1)^{\sum_{\substack{j<k \\ \pi(j)>\pi(k)}} \sigma(a^{\pi(j)})\sigma(a^{\pi(k)})}.$$

An application of (6.16) shows that (6.24) is equal to

$$\sum_{i=1}^{m} \left(\sum_{\pi \in S_m} Y_\pi a^{\pi(1)} \otimes \cdots \otimes a^{\pi(m)} \right)^{M_m} \bullet^{\{i\}} b. \tag{6.25}$$

For our purposes, (6.25) is better re-written as

$$\sum_{\pi \in S_m} Y_\pi \sum_{i=1}^{m} (a^{\pi(1)} \otimes \cdots \otimes a^{\pi(m)})^{M_m} \bullet^{\{\pi^{-1}(i)\}} b$$

which by (6.5) is equal to

$$\sum_{\pi \in S_m} Y_\pi \sum_{i=1}^{m} (-1)^{\sum_{k > \pi^{-1}(i)} \sigma(b)\sigma(a^{\pi(k)})} a^{\pi(1)} \otimes \cdots \otimes a^i b \otimes \cdots \otimes a^{\pi(m)}. \tag{6.26}$$

Now, given an arbitrary permutation $\pi \in S_m$, the expression

$$(-1)^{\sum_{\substack{\pi^{-1}(i)<k \\ i>\pi(k)}} \sigma(b)\sigma(a^{\pi(k)}) + \sum_{\substack{k<\pi^{-1}(i) \\ \pi(k)>i}} \sigma(b)\sigma(a^{\pi(k)}) + \sum_{k>\pi^{-1}(i)} \sigma(b)\sigma(a^{\pi(k)})}$$

$$\tag{6.27}$$

cancels down to

$$(-1)^{\sum_{\substack{\pi^{-1}(i)<k \\ i\le\pi(k)}} \sigma(b)\sigma(a^{\pi(k)})+\sum_{\substack{k<\pi^{-1}(i) \\ \pi(k)>i}} \sigma(b)\sigma(a^{\pi(k)})} . \tag{6.28}$$

Consider the clause $i \le \pi(k)$ in the left-hand sum. If $i = \pi(k)$ then $\pi^{-1}(i) = k$ which is disallowed by the clause $\pi^{-1}(i) < k$. Thus (6.28) is equal to

$$(-1)^{\sum_{\substack{\pi^{-1}(i)<k \\ i<\pi(k)}} \sigma(b)\sigma(a^{\pi(k)})+\sum_{\substack{k<\pi^{-1}(i) \\ \pi(k)>i}} \sigma(b)\sigma(a^{\pi(k)})} . \tag{6.29}$$

It is easy to see that (6.29) is equal to

$$(-1)^{\sum_{i<\pi(k)} \sigma(b)\sigma(a^{\pi(k)})}$$

which is simply

$$(-1)^{\sum_{i<k} \sigma(b)\sigma(a^k)} . \tag{6.30}$$

We have now established the relation

$$(-1)^{\sum_{\substack{\pi^{-1}(i)<k \\ i>\pi(k)}} \sigma(b)\sigma(a^{\pi(k)})+\sum_{\substack{k<\pi^{-1}(i) \\ \pi(k)>i}} \sigma(b)\sigma(a^{\pi(k)})+\sum_{k>\pi^{-1}(i)} \sigma(b)\sigma(a^{\pi(k)})}$$
$$=(-1)^{\sum_{i<k} \sigma(b)\sigma(a^k)}$$

and so may deduce by simple manipulation that

$$(-1)^{\sum_{k>\pi^{-1}(i)} \sigma(b)\sigma(a^{\pi(k)})}$$
$$=(-1)^{\sum_{i<k} \sigma(b)\sigma(a^k)+\sum_{\substack{\pi^{-1}(i)<k \\ i>\pi(k)}} \sigma(b)\sigma(a^{\pi(k)})+\sum_{\substack{k<\pi^{-1}(i) \\ \pi(k)>i}} \sigma(b)\sigma(a^{\pi(k)})} . \tag{6.31}$$

Note that the power of -1 taken by the term in $\imath([a^1 \otimes \cdots \otimes a^i b \otimes \cdots \otimes a^n])$ corresponding to an arbitrary permutation π is

$$Y_\pi(-1)^{\sum_{\substack{\pi^{-1}(i)<k \\ i>\pi(k)}} \sigma(b)\sigma(a^{\pi(k)})+\sum_{\substack{k<\pi^{-1}(i) \\ \pi(k)>i}} \sigma(b)\sigma(a^{\pi(k)})} . \tag{6.32}$$

For convenience we define $\tilde{Y}_i$ for each $i = 1,\ldots,m$ by

$$\tilde{Y}_i = (-1)^{\sum_{i<k} \sigma(b)\sigma(a^k)}.$$

Armed with (6.31) and this new item of notation we may re-write (6.26) as

$$\left(\sum_{i=1}^m \tilde{Y}_i \sum_{\pi\in S_m} Y_\pi(-1)^{\sum_{\substack{\pi^{-1}(i)<k \\ i>\pi(k)}} \sigma(b)\sigma(a^{\pi(k)})+\sum_{\substack{k<\pi^{-1}(i) \\ \pi(k)>i}} \sigma(b)\sigma(a^{\pi(k)})}\right)$$
$$a^{\pi(1)} \otimes \cdots \otimes a^i b \otimes \cdots \otimes a^{\pi(n)} \tag{6.33}$$

which, by the remark just made concerning (6.32), is equal to

$$\sum_{i=1}^{m}(-1)^{\sum_{i<k}\sigma(b)\sigma(a^k)}\imath([a^1\otimes\cdots\otimes a^ib\otimes\cdots\otimes a^m])$$

which, in turn, is equal to

$$\imath([\sum_{i=1}^{m}(-1)^{\sum_{i<k}\sigma(b)\sigma(a^k)}a^1\otimes\cdots a^ib\otimes\cdots\otimes a^m])$$

which is clearly a supersymmetric tensor.

The second of the two sums constituting $\imath([a^1\otimes\cdots\otimes a^m])\star(0,b,0,0,\ldots)$ corresponds to the case where A is of the form $M_{m+1}\setminus\{i\}$ and B is of the form $\{i\}$. With Y_π defined as before, this sum is

$$\sum_{i=1}^{m+1}\left(\sum_{\pi\in S_m}Y_\pi a^{\pi(1)}\otimes\cdots\otimes a^{\pi(m)}\right)_{M_{m+1}\setminus\{i\}}\bullet_{\{i\}}b. \qquad (6.34)$$

The sum (6.34) may be re-written as

$$\sum_{i=1}^{m+1}\left(\sum_{\pi\in S_m}(-1)^{\displaystyle\sum_{\substack{j<k\\\pi^{-1}(j)>\pi^{-1}(k)}}\sigma(a^j)\sigma(a^k)}a^{\pi(1)}\otimes\cdots\otimes a^{\pi(m)}\right)_{M_{m+1}\setminus\{i\}}\bullet_{\{i\}}b.$$

$$(6.35)$$

Applying (6.5), the sum (6.35) can be seen to be equal to

$$\sum_{i=1}^{m+1}\sum_{\pi\in S_m}Y'_{(\pi,i)}a^{\pi(1)}\otimes\cdots\otimes a^{\pi(i-1)}\otimes b\otimes a^{\pi(i)}\otimes\cdots\otimes a^{\pi(m)}. \qquad (6.36)$$

Here we define for each pair (π,i) the value $Y'_{(\pi,i)}$ by

$$Y'_{(\pi,i)}=(-1)^{\displaystyle\sum_{\substack{1\le j<k\le m\\\pi^{-1}(j)>\pi^{-1}(k)}}\sigma(a^j)\sigma(a^k)+\sum_{\substack{j<m+1\\\pi^{-1}(j)>i-1}}\sigma(a^j)\sigma(b)}.$$

Set $c^i=a^i$ for each $i=1,\ldots,m$ and $c^{m+1}=b$ so that (6.36) becomes

$$\sum_{i=1}^{m+1}\sum_{\pi\in S_m}Y''_{(\pi,i)}c^{\pi(1)}\otimes\cdots\otimes c^{\pi(i-1)}\otimes c^{m+1}\otimes c^{\pi(i)}\otimes\cdots\otimes c^{\pi(m)}. \qquad (6.37)$$

Here we define for each pair (π,i) the value $Y''_{(\pi,i)}$ by

$$Y''_{(\pi,i)}=(-1)^{\displaystyle\sum_{\substack{1\le j<k\le m\\\pi^{-1}(j)>\pi^{-1}(k)}}\sigma(c^j)\sigma(c^k)+\sum_{\substack{j<m+1\\\pi^{-1}(j)>i-1}}\sigma(c^j)\sigma(c^{m+1})}.$$

Of course, for a given (π,i) we have $Y'_{(\pi,i)}=Y''_{(\pi,i)}$ but the reason for the introduction of these notations is typographical rather than mathematical.

If for each $i = 1, \ldots, m+1$ we define ρ_i to be the permutation $(i\ i+1 \ldots m\ m+1)$ so that $\rho_i^{-1} = (m+1\ m\ \ldots\ i+1\ i)$ then (6.37) may be re-written as

$$\sum_{i=1}^{m+1} \sum_{\pi \in S_m} Y''_{(\pi,i)} c^{\pi\rho_i^{-1}(1)} \otimes \cdots \otimes c^{\pi\rho_i^{-1}(i-1)} \otimes c^{\rho_i^{-1}(i)} \otimes c^{\pi\rho_i^{-1}(i+1)} \otimes \cdots \otimes c^{\pi\rho_i^{-1}(m)}.$$

(6.38)

It will now be argued that the summation $\sum_{i=1}^{m+1} \sum_{\pi \in S_m}$ may be replaced by the summation $\sum_{\nu \in S_{m+1}}$. It is clear that the permutation defined by

$$j \mapsto \begin{cases} \pi\rho_i^{-1}(j) & \text{if } j \neq i; \\ m+1 & \text{if } j = i. \end{cases}$$

is always an element of S_{m+1}. Note that $m+1 = \rho_i^{-1}(i)$ for all $i = 1, \ldots, m+1$. It is also clear that distinct values of (π, i) will yield distinct elements of S_{m+1} under this definition. We must now show for an arbitrary element ν of S_{m+1} that there exists a pair (π, i) with $\pi \in S_m$ and $i \in M_m$ such that

$$\nu = \begin{cases} \pi\rho_i^{-1} & \text{on } M_{m+1} \setminus \{i\}; \\ \rho_i^{-1} & \text{on } \{i\}. \end{cases}$$

We fix i uniquely by requiring that $\nu(i) = m+1$. We now fix the permutation π. For $j < i$ constrain π to be a permutation in S_m which maps j to $\nu(j)$. With j such that $i < j \leq m+1$ constrain π further to be a permutation in S_m which maps $j-1$ to $\nu(j)$. We have now identified a unique element π of S_m. Recalling that $\rho_i^{-1} = (m+1\ m\ \ldots\ i+1\ i)$ we have that

$$\pi\rho_i^{-1}(j) = \pi(j) = \nu(j) \text{ if } 1 < j < i;$$
$$\pi\rho_i^{-1}(j) = \pi(j-1) = \nu(j) \text{ if } i < j \leq m+1;$$
$$\rho_i^{-1}(j) = \rho_i^{-1}(i) = m+1 \text{ if } j = i.$$

Thus we have identified a pair (π, i) corresponding to ν. It is clear that this correspondence is injective and so we have established an explicit bijection between S_{m+1} and $\{(\pi, i) : \pi \in S_m,\ i \in M_{m+1}\}$ as required.

Note that if ν is the permutation in S_{m+1} corresponding to a particular pair (π, i) then $\nu = \pi\rho_i^{-1}$ on $M_{m+1} \setminus \{i\}$. It follows that $\nu\rho_i = \pi$ on M_m and so $\pi^{-1} = \rho_i^{-1}\nu^{-1}$ on M_m. Furthermore, if (π, i) is a pair with corresponding S_{m+1} permutation ν then we have that $i = \nu^{-1}(m+1)$. For convenience we define for each ν in S_{m+1} the permutation α_ν in S_{m+1} by $\alpha_\nu = \left(\rho_{\nu^{-1}(m+1)}\right)^{-1}$. Also as a matter of convenience we write for each permutation ν in S_{m+1} the tensor $c^{\nu(1)} \otimes \cdots \otimes c^{\nu(m+1)}$ as c^ν. We may now re-write (6.38) as

$$\sum_{\nu \in S_{m+1}} (-1)^{\sum_{\substack{1 \leq j < k \leq m \\ \alpha_\nu \nu^{-1}(j) > \alpha_\nu \nu^{-1}(k)}} \sigma(c^j)\sigma(c^k) + \sum_{\substack{j < m+1 \\ \alpha_\nu \nu^{-1}(j) > \nu^{-1}(m+1)-1}} \sigma(c^j)\sigma(c^{m+1})} c^\nu.$$

(6.39)

In the sum

$$\sum_{\substack{1 \le j < k \le m \\ \alpha_\nu \nu^{-1}(j) > \alpha_\nu \nu^{-1}(k)}} \sigma(c^j)\sigma(c^k)$$

of (6.39), the clause $\alpha_\nu \nu^{-1}(j) > \alpha_\nu \nu^{-1}(k)$ is always equivalent to $\nu^{-1}(j) > \nu^{-1}(k)$ as the permutation $\left(\rho_{\nu^{-1}(m+1)}\right)^{-1}$ is order-preserving for all elements of M_{m+1} except $\nu^{-1}(m+1)$ and, if $\nu^{-1}(k) = \nu^{-1}(m+1)$, then $k = m+1$ which is excluded by the clause $1 \le j < k \le m$. Similarly, the case $\nu^{-1}(j) = \nu^{-1}(m+1)$ does not arise. We can also see that in the sum

$$\sum_{\substack{j < m+1 \\ \alpha_\nu \nu^{-1}(j) > \nu^{-1}(m+1)-1}} \sigma(c^j)\sigma(c^{m+1})$$

the clause $\alpha_\nu \nu^{-1}(j) > \nu^{-1}(m+1) - 1$ is equivalent to $\nu^{-1}(j) > \nu^{-1}(m+1)$ as $\left(\rho_{\nu^{-1}(m+1)}\right)^{-1}$ decreases all values greater than $\nu^{-1}(m+1)$ by one and the case $\nu^{-1}(j) = \nu^{-1}(m+1)$ does not arise because of the $j < m+1$ clause. Thus we see that (6.39) is equal to

$$\sum_{\nu \in S_{m+1}} (-1)^{\sum_{\substack{1 \le j < k \le m \\ \nu^{-1}(j) > \nu^{-1}(k)}} \sigma(c^j)\sigma(c^k) + \sum_{\substack{j < m+1 \\ \nu^{-1}(j) > \nu^{-1}(m+1)}} \sigma(c^j)\sigma(c^{m+1})} c^\nu.$$

(6.40)

We amalgamate the two sums of grading factors in (6.40) so that, with j, k running over all of M_{m+1}, (6.40) may be re-written as

$$\sum_{\nu \in S_{m+1}} (-1)^{\sum_{\substack{j < k \\ \nu^{-1}(j) > \nu^{-1}(k)}} \sigma(c^j)\sigma(c^k)} c^{\nu(1)} \otimes \cdots \otimes c^{\nu(m+1)}.$$

(6.41)

In a similar manner to the derivation of (6.18) from (6.17), we may re-write (6.41) as

$$\sum_{\nu \in S_{m+1}} (-1)^{\sum_{\substack{j < k \\ \nu(j) > \nu(k)}} \sigma(c^{\nu(j)})\sigma(c^{\nu(k)})} c^{\nu(1)} \otimes \cdots \otimes c^{\nu(m+1)}$$

(6.42)

which by definition (6.16) is equal to

$$\imath([c^1 \otimes \cdots \otimes c^{m+1}]).$$

(6.43)

By simple application of the definition of the c^i given above, (6.43) may be re-written as

$$\imath([a^1 \otimes \cdots \otimes a^m \otimes b])$$

which is a supersymmetric tensor, as required. It is obvious that the sum of two supersymmetric tensors is itself a supersymmetric tensor and so we have that

$$\imath([a^1 \otimes \cdots \otimes a^n]) \star (0, b, 0, 0, \ldots)$$

is a supersymmetric tensor.

We now must extend this result so that it holds for arbitrary supersymmetric tensors $\imath([c])$ on the right of the product $\imath([a^1 \otimes \cdots \otimes a^m]) \star \imath([c])$. Consider the set of finite products

$$C := \{(0, a_1, 0, \ldots) \star \cdots \star (0, a_n, 0, \ldots) : n \in \mathbf{N}, \ a_1, a_2, \ldots, a_n \in \mathcal{I}\}.$$

We claim that C spans the space $\mathcal{S}(\mathcal{I})$. Indeed, $(0, a_1, 0, 0, \ldots) = \imath([a_1])$ is clearly a supersymmetric tensor. If, by way of induction, we assume that $(0, a_1, 0, \ldots) \star \cdots \star (0, a_{n-1}, 0, \ldots)$ has highest order component $\imath([a_1 \otimes \cdots \otimes a_{n-1}])$ then, by the result just proved concerning the second sum associated with the product $\imath([d^1 \otimes \cdots \otimes d^n]) \star (0, e, 0, \ldots)$ where $d^1, \ldots, d^n, e$ are arbitrary elements of $\mathcal{I}$, we have that the product

$$(0, a_1, 0, \ldots) \star \cdots \star (0, a_{n-1}, 0, \ldots) \star (0, a_n, 0, \ldots)$$

has highest order component $\imath([a^1 \otimes \cdots \otimes a^{n-1} \otimes a^n])$. It follows, then, that the elements of C span $\mathcal{S}(\mathcal{I})$. Thus we have $a \star b \in \mathcal{S}_S(\mathcal{I})$ for arbitrary $a \in \mathcal{S}_S(\mathcal{I})$ and arbitrary $b \in \mathcal{S}(\mathcal{I})$. For arbitrary $b \in \mathcal{S}_S(\mathcal{I})$, the nth component $(a \star b)_n$ of $a \star b$ may be formed by replacing b with the truncation of b at the nth entry. This means that $a \star b \in \mathcal{S}_S(\mathcal{I})$ for arbitrary $a, b \in \mathcal{S}_S(\mathcal{I})$ as required.∎

Corollary 6.5 Under the multiplication $\star$, the supersymmetric subalgebra $\mathcal{S}(\mathcal{I})$ defined above is a unital sub-†-superalgebra of $\mathcal{T}(\mathcal{I})$.

7 Some Results in $\mathbf{Z}_2$-Graded Quantum Stochastic Calculus

Throughout this chapter we consider N and r to be fixed natural numbers where $N \geq 1$ and $0 \leq r < N$.

7.1 The First Fundamental Formula in $\mathbf{Z}_2$-Graded Quantum Stochastic Calculus

Consider an ungraded quantum stochastic integral of the form

$$\int_0^t K(s)\, d\Lambda_\beta^\alpha(s)$$

where $t \geq 0$, $0 \leq \alpha, \beta \leq N$ and K is a process integrable by $d\Lambda_\beta^\alpha$. The first fundamental formula of quantum stochastic calculus given in theorem 2.7 states that, for arbitrary exponential vectors $e(f), e(g)$ with $f, g \in \mathrm{L}^2(\mathbf{R}_+; \mathbf{C}^N)$ we have

$$\langle e(f), \int_0^t K(s)\, d\Lambda_\beta^\alpha(s) e(g)\rangle = \int_0^t f_\beta(s) g^\alpha(s) \langle e(f), K(s) e(g)\rangle\, ds. \qquad (7.1)$$

From (7.1) we may derive an equivalent formula for $\mathbf{Z}_2$-graded quantum stochastic calculus. Again, take arbitrary α, β with $0 \leq \alpha, \beta \leq N$, arbitrary $t \geq 0$ and an arbitrary process K such that the quantum stochastic integral

$$\int_0^t K(s)\, d\Xi_\beta^\alpha(s) \qquad (7.2)$$

exists. The integral (7.2) is, from the definition of the differential $d\Xi_\beta^\alpha$ given by (4.16), the same as

$$\int_0^t K(s) G^{\sigma_\beta^\alpha}(s)\, d\Lambda_\beta^\alpha(s). \qquad (7.3)$$

Putting (7.3) into (7.1) gives

$$\langle e(f), \int_0^t K(s)G^{\sigma\tilde{\alpha}}_{\beta}(s)\,d\Lambda^{\alpha}_{\beta}(s)e(g)\rangle =$$

$$\int_0^t f_{\beta}(s)g^{\alpha}(s)\langle e(f), K(s)G^{\sigma\tilde{\alpha}}_{\beta}(s)e(g)\rangle\,ds. \tag{7.4}$$

Thus, from (7.4), we have the following simple form of the first fundamental formula in the $\mathbf{Z}_2$-graded case:

$$\langle e(f), \int_0^t K(s)\,d\Xi^{\alpha}_{\beta}(s)e(g)\rangle = \int_0^t f_{\beta}(s)g^{\alpha}(s)\langle e(f), K(s)G^{\sigma\tilde{\alpha}}_{\beta}(s)e(g)\rangle\,ds. \tag{7.5}$$

The generalisation of (7.5) to a differential $d\Xi_A$ where A is an arbitrary element of $M_0(N, r)$ which decomposes as $\lambda^{\beta}_{\alpha}E^{\alpha}_{\beta}$ with each $\lambda^{\beta}_{\alpha} \in \mathbf{C}$ is

$$\langle e(f), \int_0^t K(s)\,d\Xi_A e(g)\rangle = \sum_{\alpha,\beta=0}^{N} \int_0^t \lambda^{\beta}_{\alpha} f_{\beta}(s)g^{\alpha}(s)\langle e(f), K(s)G^{\sigma\tilde{\alpha}}_{\beta}(s)e(g)\rangle\,ds. \tag{7.6}$$

This might be further generalised to to integrals of the form

$$\int_0^t F^{\beta}_{\alpha}(s)\,d\Xi^{\alpha}_{\beta}(s)$$

where each F^{β}_{α} is integrable with respect to the appropriate $d\Xi^{\alpha}_{\beta}$. In this case we have

$$\langle e(f), \int_0^t F^{\beta}_{\alpha}(s)\,d\Xi^{\alpha}_{\beta}(s)e(g)\rangle =$$

$$\sum_{\alpha,\beta=0}^{N} \int_0^t f_{\beta}(s)g^{\alpha}(s)\langle e(f), F^{\beta}_{\alpha}(s)G^{\sigma\tilde{\alpha}}_{\beta}(s)e(g)\rangle\,ds. \tag{7.7}$$

7.2 The Second Fundamental Formula in $\mathbf{Z}_2$-Graded Quantum Stochastic Calculus

We now consider the second fundamental formula introduced in [HP1] and stated in theorem 2.8. We begin by taking arbitrary $\alpha, \beta, \gamma, \delta$ with $0 \leq \alpha, \beta, \gamma, \delta \leq N$, arbitrary exponential vectors $e(f), e(g)$ with f, g in $L^2(\mathbf{R}_+; \mathbf{C}^N)$, arbitrary $t \geq 0$ and considering the following inner product in which the integrands are both the identity process:

$$\langle \int_0^t d\Xi^{\beta}_{\alpha}(s)e(f), \int_0^t d\Xi^{\gamma}_{\delta}(s)e(g)\rangle. \tag{7.8}$$

The ungraded version of the second fundamental formula and the definition of the $d\Xi^{\epsilon}_{\mu}$ given in (4.16) shows that (7.8) is equal to

$$\int_0^t f_\delta(s)g^\gamma(s)\langle \int_0^s d\Xi_\alpha^\beta(u)e(f), G^{\sigma\gamma}_\delta(s)e(g)\rangle\, ds$$

$$+\int_0^t f_\beta(s)g^\alpha(s)\langle G^{\sigma\beta}_\alpha(s)e(f), \int_0^s d\Xi_\delta^\gamma(u)e(g)\rangle\, ds \qquad (7.9)$$

$$+\hat{\delta}_\delta^\alpha \int_0^t f_\beta(s)g^\gamma(s)\langle G^{\sigma\beta}_\alpha(s)e(f), G^{\sigma\gamma}_\delta(s)e(g)\rangle\, ds.$$

This result was previously seen in Chapter 4.

If K and L are quantum stochastic processes integrable by $d\Xi_\alpha^\beta$ and $d\Xi_\delta^\gamma$ respectively and $e(f), e(g)$ are exponential vectors with f, g being arbitrary elements of $\mathbf{L}^2(\mathbf{R}_+; \mathbf{C}^N)$ then the inner product

$$\langle \int_0^t K(s)\, d\Xi_\alpha^\beta(s)e(f), \int_0^t L(s)\, d\Xi_\delta^\gamma(s)e(g)\rangle$$

may be re-written in the language of ungraded quantum stochastic calculus as

$$\langle \int_0^t K(s)G^{\sigma\beta}_\alpha(s)\, d\Lambda_\alpha^\beta(s)e(f), \int_0^t L(s)G^{\sigma\gamma}_\delta(s)\, d\Lambda_\delta^\gamma(s)e(g)\rangle. \qquad (7.10)$$

The second fundamental formula shows that (7.10) is equal to

$$\int_0^t f_\delta(s)g^\gamma(s)\langle \int_0^s K(u)G^{\sigma\beta}_\alpha(u)d\Lambda_\alpha^\beta(u)e(f), L(s)G^{\sigma\gamma}_\delta(s)e(g)\rangle\, ds$$

$$+\int_0^t f_\beta(s)g^\alpha(s)\langle K(s)G^{\sigma\beta}_\alpha(s)e(f), \int_0^s L(u)G^{\sigma\gamma}_\delta(u)\, d\Lambda_\delta^\gamma(u)e(g)\rangle\, ds \qquad (7.11)$$

$$+\hat{\delta}_\delta^\alpha \int_0^t f_\beta(s)g^\gamma(s)\langle K(s)G^{\sigma\beta}_\alpha(s)e(f), L(s)G^{\sigma\gamma}_\delta(s)e(g)\rangle\, ds.$$

Translated into the language of $\mathbf{Z}_2$-graded quantum stochastic calculus, (7.11) becomes

$$\int_0^t f_\delta(s)g^\gamma(s)\langle \int_0^s K(u)\, d\Xi_\alpha^\beta(u)e(f), L(s)G^{\sigma\gamma}_\delta(s)e(g)\rangle\, ds$$

$$+\int_0^t f_\beta(s)g^\alpha(s)\langle K(s)G^{\sigma\beta}_\alpha(s)e(f), \int_0^s L(u)\, d\Xi_\delta^\gamma(u)e(g)\rangle\, ds \qquad (7.12)$$

$$+\hat{\delta}_\delta^\alpha \int_0^t f_\beta(s)g^\gamma(s)\langle K(s)G^{\sigma\beta}_\alpha(s)e(f), L(s)G^{\sigma\gamma}_\delta(s)e(g)\rangle\, ds.$$

We now offer an extension of (7.12). Consider a pair of general integrals of the form

$$\int_0^t K_\beta^\alpha(s)\, d\Xi_\alpha^\beta(s), \qquad \int_0^t L_\gamma^\delta(s)\, d\Xi_\delta^\gamma(s)$$

where we assume each K_β^α is integrable against its corresponding $d\Xi_\alpha^\beta$ and each L_γ^δ is integrable against its corresponding $d\Xi_\delta^\gamma$. We may extend (7.12) to these more general integrals as follows:

$$\langle \int_0^t K_\beta^\alpha(s)\,d\Xi_\alpha^\beta(s)e(f), \int_0^t L_\gamma^\delta(s)\,d\Xi_\delta^\gamma(s)e(g)\rangle$$

$$= \sum_{\gamma,\delta=0}^N \int_0^t f_\delta(s)g^\gamma(s)\langle \int_0^s K_\beta^\alpha(s)\,d\Xi_\alpha^\beta(s)e(f), L_\gamma^\delta(s)G^{\sigma\gamma}_\delta(s)e(g)\rangle$$

$$+ \sum_{\alpha,\beta=0}^N \int_0^t f_\beta(s)g^\alpha(s)\langle K_\beta^\alpha(s)G^{\sigma\beta}_\alpha(s)e(f), \int_0^s L_\gamma^\delta(s)\,d\Xi_\delta^\gamma(u)e(g)\rangle\,ds \qquad (7.13)$$

$$+ \sum_{\beta,\gamma=0}^N \sum_{i=1}^N \int_0^t f_\beta(s)g^\gamma(s)\langle K_\beta^i(s)G^{\sigma\beta}_i(s)e(f), L_\gamma^i(s)G^{\sigma\gamma}_i(s)e(g)\rangle\,ds.$$

While (7.13) is a more general formulation of the second fundamental formula for the Z_2-graded case, it will prove more convenient to work with (7.12). In the next section we use the second fundamental formula to improve our understanding of the nature of Z_2-graded quantum stochastic integrals.

7.3 Discussion of the Second Fundamental Formula in Z_2-Graded Quantum Stochastic Calculus

In this section we use a formal version of the second fundamental formula to provide some insights into the nature of Z_2-graded quantum stochastic integrals. We begin by determining a preliminary form for the formal second fundamental formula in the Z_2-graded case. This is followed by a discussion of the means by which quantum stochastic processes, including quantum stochastic integrals, are graded. The results obtained here enable us to re-work the preliminary form of the second fundamental formula into a final version which we will use to clarify the role of the Chevalley tensor product in Z_2-graded quantum stochastic calculus. We end the section with a discussion of the differential form of the version of the second fundamental formula that we have derived.

For the moment we retain the inner product but take adjoints of the unbounded operators on the left of the inner products in (7.12). By definition (4.16) and the standard results discussed in Chapter 2, the adjoint of the integral

$$\int_0^s K(u)\,d\Xi_\alpha^\beta(u)$$

is

$$\left(\int_0^s K(u)G^{\sigma\beta}_\alpha(u)\,d\Lambda_\alpha^\beta(u)\right)^\dagger$$

$$= \int_0^s G^{\sigma\beta}_\beta(u)K^\dagger(u)\,d\Lambda_\beta^\alpha(u).$$

We may now re-write (7.12) formally as

$$\int_0^t f_\delta(s)g^\gamma(s)\langle e(f), \int_0^s G^{\sigma^a_\beta}(u)K^\dagger(u)\,d\Lambda^\alpha_\beta(u)L(s)G^{\sigma^\gamma_\delta}(s)e(g)\rangle\,ds$$

$$+\int_0^t f_\beta(s)g^\alpha(s)\langle e(f), G^{\sigma^a_\beta}(s)K^\dagger(s)\int_0^s L(u)\,d\Xi^\gamma_\delta(u)e(g)\rangle\,ds \qquad (7.14)$$

$$+\hat{\delta}^\alpha_\delta\int_0^t f_\beta(s)g^\gamma(s)\langle e(f), G^{\sigma^a_\beta}(s)K^\dagger(s)L(s)G^{\sigma^\gamma_\delta}(s)e(g)\rangle\,ds.$$

Applying the ungraded version of the first fundamental formula given in theorem 2.7 to (7.14) yields

$$\langle e(f), \int_0^t\int_0^s G^{\sigma^a_\beta}(u)K^\dagger(u)\,d\Lambda^\alpha_\beta(u)L(s)G^{\sigma^\gamma_\delta}(s)\,d\Lambda^\gamma_\delta(s)e(g)\rangle$$

$$+\langle e(f), \int_0^t G^{\sigma^a_\beta}(s)K^\dagger(s)\int_0^s L(u)\,d\Xi^\gamma_\delta(u)\,d\Lambda^\alpha_\beta e(g)\rangle \qquad (7.15)$$

$$+\hat{\delta}^\alpha_\delta\langle e(f), \int_0^t G^{\sigma^a_\beta}(s)K^\dagger(s)L(s)G^{\sigma^\gamma_\delta}(s)\,d\Lambda^\gamma_\beta e(g)\rangle.$$

We now dispense with the inner product and write formally

$$\left(\int_0^t G^{\sigma^a_\beta}(s)K^\dagger(s)\,d\Lambda^\alpha_\beta(s)\right)\left(\int_0^t L(s)\,d\Xi^\gamma_\delta(s)\right)$$

$$=\int_0^t\int_0^s G^{\sigma^a_\beta}(u)K^\dagger(u)\,d\Lambda^\alpha_\beta(u)L(s)\,d\Xi^\gamma_\delta(s)$$

$$+\int_0^t G^{\sigma^a_\beta}(s)K^\dagger(s)\int_0^s L(u)\,d\Xi^\gamma_\delta(u)\,d\Lambda^\alpha_\beta(s) \qquad (7.16)$$

$$+\hat{\delta}^\alpha_\delta\int_0^t G^{\sigma^a_\beta}(s)K^\dagger(s)L(s)G^{\sigma^\gamma_\delta}(s)\,d\Lambda^\gamma_\beta(s).$$

It was seen in Section 4.2 that $\Gamma(L^2(\mathbf{R}_+; \mathbf{C}))$ is graded by each $G(s)$ as $\Gamma_{s0}\oplus\Gamma_{s1}$. Here Γ_{s0} corresponds to the eigenspace of the eigenvalue 1 of the self-adjoint unitary operator $G(s)$ acting on the Fock space and Γ_{s1} corresponds to the eigenspace of the eigenvalue -1 of $G(s)$ acting on the Fock space. In a similar way, the space of linear operators acting in $\Gamma(L^2(\mathbf{R}_+; \mathbf{C}^N))$ can be graded by G. To do this we use the operator $G(\infty)$ defined on the whole Fock space by its action on an arbitrary exponential vector $e(f)$ where $f = (f^1, \ldots, f^N)\in L^2(\mathbf{R}_+, \mathbf{C}^N)$ as follows:

$$G(\infty)e(f) = e((f^1, \ldots, f^r, -f^{r+1}, \ldots, -f^N)).$$

An operator X is said to be of parity 0 if $G(\infty)XG(\infty) = X$ and said to be of parity 1 if $G(\infty)XG(\infty) = -X$. As $G(\infty)$ is a self-adjoint unitary, any linear operator Y in Fock space may be decomposed as $Y_0 + Y_1$ so that

$$G(\infty)YG(\infty) = G(\infty)(Y_0 + Y_1)G(\infty) = Y_0 - Y_1. \tag{7.17}$$

It is, in fact, more appropriate in this work to grade the processes consisting of operators on the Fock space $\Gamma(L^2(\mathbf{R}_+; \mathbf{C}^N))$. A process $X = (X(t))_{t \geq 0}$ is said to be of parity 0 if, for all $t \geq 0$, $G(\infty)X(t)G(\infty) = X(t)$. If, on the other hand, for all $t \geq 0$ we have $G(\infty)X(t)G(\infty) = -X(t)$ then we say that X is of parity 1. It is clear that, as with operators, any process can be expressed as the sum of two processes of definite parity. Note that in expressions such as $\int_0^t \langle e(f), G(s)e(g) \rangle \, ds$, the $G(s)$ might equally well be replaced by $G(\infty)$.

Let us now return to the formal version of the second fundamental formula. We assume that the processes L and K are of definite parity. For each $s \geq 0$, we have

$$\begin{aligned} G(\infty)K^\dagger(s)G(\infty) &= (G(\infty)K(s)G(\infty))^\dagger \\ &= ((-1)^{\sigma(K)}K(s))^\dagger = (-1)^{\sigma(K)}K^\dagger(s). \end{aligned} \tag{7.18}$$

We conclude from (7.18) that $\sigma(M^\dagger) = \sigma(M)$ for any process M of definite parity. These observations and (4.16) allow us to re-write equation (7.16) as

$$\begin{aligned} (-1)^{\sigma_\beta^\alpha \sigma(K)} &\left(\int_0^t K^\dagger(s) \, d\Xi_\beta^\alpha(s) \right) \left(\int_0^t L(s) \, d\Xi_\delta^\gamma(s) \right) \\ &= (-1)^{\sigma_\beta^\alpha \sigma(K)} \int_0^t \int_0^s K^\dagger(u) \, d\Xi_\beta^\alpha(u) L(s) \, d\Xi_\delta^\gamma(s) \\ &\quad + (-1)^{\sigma_\beta^\alpha \sigma(K) + \sigma_\beta^\alpha \sigma\left(\int_0^s L(u) \, d\Xi_\delta^\gamma(u) \right)} \int_0^t K^\dagger(s) \int_0^s L(u) \, d\Xi_\delta^\gamma(u) \, d\Xi_\beta^\alpha(s) \\ &\quad + (-1)^{\sigma_\beta^\alpha(\sigma(K) + \sigma(L))} \hat{\delta}_\delta^\alpha \int_0^t K^\dagger(s) L(s) G^{\sigma_\beta^\alpha + \sigma_\delta^\gamma}(s) \, d\Lambda_\beta^\gamma(s). \end{aligned} \tag{7.19}$$

The second term on the right of (7.19) requires some discussion. While $\int_0^s L(u) \, d\Xi_\delta^\gamma(u)$ is certainly a quantum stochastic process, we have not established that it is of definite parity or what this parity might be. In order that we might proceed we must state and prove a proposition.

Proposition 7.1 Let A be an arbitrary homogeneous element of $M_0(N, r)$ and let M be an arbitrary quantum stochastic process of definite parity. Then for arbitrary $t \geq 0$ the integral

$$\int_0^t M(s) \, d\Xi_A(s)$$

is of definite parity and this parity is equal to $\sigma(A) + \sigma(M) \pmod 2$.

Proof. By linearity it suffices to consider the case where A is a basis element E_β^α of $M_0(N, r)$. Consider the inner product

$$\langle e(f), G(\infty) \int_0^t M(s)\, d\Xi_\beta^\alpha(s) G(\infty) e(g) \rangle \qquad (7.20)$$

where $e(f), e(g)$ are exponential vectors with f, g being arbitrary elements of $L^2(\mathbf{R}_+; \mathbf{C}^N)$. The fact that $G(\infty)$ is self-adjoint allows us to re-write (7.20) as

$$\langle G(\infty) e(f), \int_0^t M(s)\, d\Xi_\beta^\alpha(s) G(\infty) e(g) \rangle. \qquad (7.21)$$

The version of the first fundamental formula given in (7.5) shows that (7.21) is equal to

$$(-1)^{\sigma_0^{\tilde\alpha} + \sigma_\beta^0} \int_0^t f_\beta(s) g^\alpha(s) \langle G(\infty) e(f), M(s) G^{\sigma_\beta^{\tilde\alpha}}(s) G(\infty) e(g) \rangle. \qquad (7.22)$$

We know by corollary 4.2 that $(-1)^{\sigma_0^{\tilde\alpha} + \sigma_\beta^0} = \sigma_\beta^\alpha \pmod 2$ and, again using the self-adjoint property of $G(\infty)$, we may move the $G(\infty)$ back to the right hand-side of the inner product. If $G(\infty)$ is then 'supercommuted' past the $M(s)$ we obtain that (7.22) is equal to

$$(-1)^{\sigma_\beta^{\tilde\alpha} + \sigma(M)} \int_0^t f_\beta(s) g^\alpha(s) \langle e(f), M(s) G(\infty) G^{\sigma_\beta^{\tilde\alpha}}(s) G(\infty) e(g) \rangle\, ds. \qquad (7.23)$$

By lemma 4.3 and the self-adjoint unitarity of $G(\infty)$ we have that (7.23) is equal to

$$(-1)^{\sigma_\beta^{\tilde\alpha} + \sigma(M)} \int_0^t f_\beta(s) g^\alpha(s) \langle e(f), M(s) G^{\sigma_\beta^{\tilde\alpha}}(s) e(g) \rangle\, ds. \qquad (7.24)$$

An application of the first fundamental formula as given in theorem 2.7 along with an application of (4.16) shows that (7.24) is equal to

$$(-1)^{\sigma_\beta^{\tilde\alpha} + \sigma(M)} \langle e(f), \int_0^t M(s)\, d\Xi_\beta^\alpha(s) e(g) \rangle \qquad (7.25)$$

so we may conclude that $\sigma \left(\int_0^t M(s)\, d\Xi_\beta^\alpha(s) \right) = \sigma_\beta^\alpha + \sigma(M)$. The required result now follows.∎

Having established this proposition, we return to (7.19). By (4.16), corollary 4.2 and (4.41) we have that the third term on the right hand side of (7.19) is equal to

$$(-1)^{\sigma_\beta^{\tilde\alpha}(\sigma(K) + \sigma(L))} \int_0^t K^\dagger(s) L(s)\, d\Xi_\beta^\alpha . d\Xi_\delta^\gamma(s). \qquad (7.26)$$

Replacing the third term on the right of (7.19) by (7.26) and implementing proposition 7.1 shows that (7.19) may be re-written as

$$(-1)^{\sigma^\alpha_\beta \sigma(K)} \left(\int_0^t K^\dagger(s)\, d\Xi^\alpha_\beta(s) \right) \left(\int_0^t L(s)\, d\Xi^\gamma_\delta(s) \right)$$

$$= (-1)^{\sigma^\alpha_\beta \sigma(K)} \int_0^t \int_0^s K^\dagger(u)\, d\Xi^\alpha_\beta(u) L(s)\, d\Xi^\gamma_\delta(s)$$

$$+ (-1)^{\sigma^\alpha_\beta \sigma(K) + \sigma^\alpha_\beta(\sigma(L) + \sigma^\gamma_\delta)} \int_0^t K^\dagger(s) \int_0^s L(u)\, d\Xi^\gamma_\delta(u)\, d\Xi^\alpha_\beta(s)$$

$$+ (-1)^{\sigma^\alpha_\beta(\sigma(K) + \sigma(L))} \int_0^t K^\dagger(s) L(s) d\Xi^\alpha_\beta . d\Xi^\gamma_\delta(s).$$

(7.27)

Cancelling the factor of $(-1)^{\sigma^\alpha_\beta \sigma(K)}$ and relabelling $K^\dagger$ as K we may re-write (7.27) as

$$\int_0^t K(s)\, d\Xi^\alpha_\beta(s) \int_0^t L(s)\, d\Xi^\gamma_\delta(s)$$

$$= \int_0^t \int_0^s K(u)\, d\Xi^\alpha_\beta(u) L(s)\, d\Xi^\gamma_\delta(s)$$

$$+ (-1)^{\sigma^\alpha_\beta(\sigma(L) + \sigma^\gamma_\delta)} \int_0^t K(s) \int_0^s L(u)\, d\Xi^\gamma_\delta(u)\, d\Xi^\alpha_\beta(s)$$

$$+ (-1)^{\sigma^\alpha_\beta \sigma(L)} \int_0^t K(s) L(s)\, d\Xi^\alpha_\beta . d\Xi^\gamma_\delta(s).$$

(7.28)

This is a convenient formal form of the second fundamental formula showing how two quantum stochastic integrals would be multiplied were this permissible.

Equation (7.28) provides some useful insights into the nature of Z_2-graded quantum stochastic integrals. In the ungraded calculus, a formal product

$$\int M\, d\Lambda_A \int N\, d\Lambda_B$$

(7.29)

can be re-written formally by the second fundamental formula given in theorem 2.8 as

$$\int \int M\, d\Lambda_A N\, d\Lambda_B + \int M \int N\, d\Lambda_B d\Lambda_A + \int_0^t MN\, d\Lambda_A . d\Lambda_B.$$

(7.30)

The third term of (7.30) is the Ito correction term. In the strict notation seen in Chapter 2, the product (7.29) should be written as

$$\int M \otimes d\Lambda_A \int N \otimes d\Lambda_B$$

with the ungraded tensor product in force. The correction term of (7.30) corresponds to

$$\int (M \otimes d\Lambda_A)(N \otimes d\Lambda_B) = \int MN \otimes d\Lambda_A . d\Lambda_B = \int MN\, d\Lambda_A . d\Lambda_B$$

using the standard tensor multiplication.

From (7.28) we see that the Ito correction term of the formal product

$$\int K \otimes d\Xi_\beta^\alpha \int L \otimes d\Xi_\delta^\gamma \tag{7.31}$$

is

$$(-1)^{\sigma_\beta^\alpha \sigma(L)} \int_0^t K(s)L(s) \otimes d\Xi_\beta^\alpha . d\Xi_\delta^\gamma(s). \tag{7.32}$$

We would like (7.32) to be equal to

$$\int (K \otimes d\Xi_\beta^\alpha)(L \otimes d\Xi_\gamma^\delta). \tag{7.33}$$

At first glance, the factor of $(-1)^{\sigma_\beta^\alpha \sigma(L)}$ in (7.32) seems to indicate that this is not the case. However, the objects under consideration are elements of graded structures and, as such, the tensor product present in the integrals of (7.31) must be the Chevalley tensor product described in Chapter 3. The general rule for the multiplication of a pair of twofold Chevalley tensors $a_1 \otimes a_2$, $b_1 \otimes b_2$ where a_1, a_2, b_1, b_2 are of definite parity was given in (3.3) as

$$(a_1 \otimes a_2)(b_1 \otimes b_2) = (-1)^{\sigma(a_2)\sigma(b_1)}(a_1 b_1 \otimes a_2 b_2).$$

Applying this rule to the product in (7.33) yields

$$(-1)^{\sigma_\beta^\alpha \sigma(L)} \int KL \otimes d\Xi_\beta^\alpha . d\Xi_\gamma^\delta$$

which is the correction term of (7.28).

In the Chevalley tensor product, a factor of -1 is introduced when, loosely speaking, one odd element passes through another in the multiplication. This heuristic idea of a factor of -1 being introduced when graded elements 'pass' one another is borne out by the iterated integrals of the right hand side of (7.28). In each of the first two terms on the right of (7.28), one of the two integrals is taken into the other to form an iterated integral. In the first term, $\int K \otimes d\Xi_\beta^\alpha$ is taken into $\int L \otimes d\Xi_\delta^\gamma$ to form $\int \left(K \otimes d\Xi_\beta^\alpha \right) L \otimes d\Xi_\delta^\gamma$. The first integral only passes through the integral sign and does not pass through any graded objects. This corresponds to the absence of a grading factor in the first term on the right hand side of (7.28). On the other hand, in order for the integral $\int L \otimes d\Xi_\delta^\gamma$ to reach its place inside $\int K \otimes d\Xi_\beta^\alpha$ to form $\int K \left(\int L \otimes d\Xi_\delta^\gamma \right) \otimes d\Xi_\beta^\alpha$, the differential $d\Xi_\beta^\alpha$ must be passed. This is a graded object and so, by proposition 7.1 and the heuristic principle in use, we would expect a factor of $(-1)^{(\sigma(L)+\sigma_\delta^\gamma)\sigma_\beta^\alpha}$ to be introduced. This is indeed the case.

We now consider the differential form of (7.28). Define the process $\tilde{K}$ to be $\int_0^\cdot K(s) \otimes d\Xi_\beta^\alpha(s)$ and the process $\tilde{L}$ to be $\int_0^\cdot L(s) \otimes d\Xi_\delta^\gamma(s)$. The differential form of (7.28) is then

$$d(\tilde{K}\tilde{L}) = \tilde{K}L \otimes d\Xi_\delta^\gamma + (-1)^{\sigma_\beta^\alpha \sigma(\tilde{L})} K\tilde{L} \otimes d\Xi_\beta^\alpha + (-1)^{\sigma_\beta^\alpha \sigma(L)} KL \otimes d\Xi_\beta^\alpha.d\Xi_\delta^\gamma. \tag{7.34}$$

Take $\mathcal{I}$ to be the superalgebra of quantum stochastic differentials $\{d\Xi_A : A \in M_0(N, r)\}$ and let $\mathcal{B}$ denote the Z_2-graded vector space of quantum stochastic processes integrable by the elements of $\mathcal{I}$. We now define an action $\Diamond$ of $\mathcal{B}$ on $\mathcal{B} \otimes \mathcal{I}$. For $B, M \in \mathcal{B}$ of definite parity and $d\Xi \in \mathcal{I}$ of definite parity we define the left and right actions of $\Diamond$ respectively by

$$B\Diamond(M \otimes d\Xi) = BM \otimes d\Xi,$$
$$(M \otimes d\Xi)\Diamond B = (-1)^{\sigma(B)\sigma(d\Xi)} MB \otimes d\Xi. \tag{7.35}$$

Note that boundedness considerations force these definitions to be considered formally. We also note that the definitions given in (7.35) agree with the comments concerning grading factors made in the previous paragraph. The action $\Diamond$ along with the definition of the product of a pair of Chevalley tensors given by (3.3) allows (7.34) to be re-written as

$$d(\tilde{K}\tilde{L}) = \tilde{K}\Diamond(L \otimes d\Xi_\delta^\gamma) + (K \otimes d\Xi_\beta^\alpha)\Diamond\tilde{L} + (K \otimes d\Xi_\beta^\alpha)(L \otimes d\Xi_\delta^\gamma). \tag{7.36}$$

This convenient, albeit strictly formal, expression of the rigorous equality (7.12) shows clearly the importance of the graded algebraic theory that underlies Z_2-graded quantum stochastic calculus. Earlier work on the relationship between Z_2-graded algebraic structures and quantum stochastic calculus may be found in [AH].

7.4 Adjoints of Z_2-Graded Quantum Stochastic Integrals

As seen in the previous section, for arbitrary α, β with $0 \le \alpha, \beta \le N$, an arbitrary time $t \ge 0$ and an arbitrary process K of definite parity which is integrable by $d\Xi_\beta^\alpha$ we have

$$\left(\int_0^t K(s) \otimes d\Xi_\beta^\alpha(s) \right)^\dagger$$

$$= \left(\int_0^t K(s)G^{\sigma_\beta^\alpha}(s) \otimes d\Lambda_\beta^\alpha(s) \right)^\dagger$$

$$= \int_0^t G^{\sigma_\beta^\alpha}(s)K^\dagger(s) \otimes d\Lambda_\alpha^\beta(s) \tag{7.37}$$

$$= (-1)^{\sigma_\beta^\alpha \sigma(K)} \int_0^t K^\dagger(s)G^{\sigma_\beta^\alpha}(s) \otimes d\Lambda_\alpha^\beta(s)$$

$$= (-1)^{\sigma_\beta^\alpha \sigma(K)} \int_0^t K^\dagger(s) \otimes d\Xi_\alpha^\beta.$$

Our aim in this section is to produce a formula for the adjoint of an iterated quantum stochastic integral of the form

$$\int_{0<t_1<\cdots<t_n<t} d\Xi_{\beta_1}^{\alpha_1}(t_1)\cdots d\Xi_{\beta_n}^{\alpha_n}(t_n). \tag{7.38}$$

It will be seen that this adjoint corresponds exactly to the involution † given in proposition 6.1. This would be expected from the fact that the tensor products implicit in (7.38) are Chevalley tensor products. First it is necessary to prove a commutation result concerning G and iterated integrals.

Proposition 7.2 For arbitrary $n \geq 1$, arbitrary α_i, β_i with $0 \leq \alpha_i, \beta_i \leq N$, $i = 1, \ldots, n$ and arbitrary $s, t \geq 0$ with $s \geq t$ (including the case $s = \infty$) we have

$$G(s) \int_{0<t_1<\cdots<t_n<t} d\Xi_{\beta_1}^{\alpha_1}(t_1)\cdots d\Xi_{\beta_n}^{\alpha_n}(t_n) =$$

$$(-1)^{\sum_{i=1}^{n} \sigma_{\beta_i}^{\alpha_i}} \int_{0<t_1<\cdots<t_n<t} d\Xi_{\beta_1}^{\alpha_1}(t_1)\cdots d\Xi_{\beta_n}^{\alpha_n}(t_n)G(s).$$

Note that this result is also to be expected from the Chevalley tensor products implicit in (7.38).

Proof. Lemma 4.4 along with (4.17) shows that the proposition holds in the case $n = 1$. Now suppose, by way of induction, that the propositions holds for all $n < k$ where k is some integer. Now take $n = k$. If $e(f), e(g)$ denote exponential vectors with f, g being arbitrary elements of $L^2(\mathbf{R}_+; \mathbf{C}^N)$ then we have from the self-adjointness of $G(s)$ that the inner product

$$\langle e(f), G(s) \int_{0<t_1<\cdots<t_k<t} d\Xi_{\beta_1}^{\alpha_1}(t_1)\cdots d\Xi_{\beta_k}^{\alpha_k}(t_k)e(g)\rangle \tag{7.39}$$

is equal to

$$\langle G(s)e(f), \int_{0<t_1<\cdots<t_k<t} d\Xi_{\beta_1}^{\alpha_1}(t_1)\cdots d\Xi_{\beta_k}^{\alpha_k}(t_k)e(g)\rangle. \tag{7.40}$$

By the first fundamental formula given in theorem 2.7 this is equal to

$$\int_0^t (-1)^{\sigma_{\beta_k}^0} f_{\beta_k}(t_k)g^{\alpha_k}(t_k)\langle G(s)e(f),$$

$$\int_{0<t_1<\cdots<t_{k-1}<t_k} d\Xi_{\beta_1}^{\alpha_1}(t_1)\cdots d\Xi_{\beta_{k-1}}^{\alpha_{k-1}}(t_{k-1})e(g)\rangle \, dt_k$$

which, by the self-adjointness of $G(s)$ and the inductive hypothesis is equal to

$$(-1)^{\sum_{i=1}^{k-1} \sigma_{\beta_i}^{\alpha_i}} \int_0^t (-1)^{\sigma_{\beta_k}^0} f_{\beta_k}(t_k) g^{\alpha_k}(t_k) \langle e(f),$$

$$\int_{0<t_1<\cdots<t_{k-1}<t_k} d\Xi_{\beta_1}^{\alpha_1}(t_1) \cdots d\Xi_{\beta_{k-1}}^{\alpha_{k-1}}(t_{k-1}) G(s)e(g) \rangle \, dt_k.$$

$$(7.41)$$

We know by corollary 4.2 that $(-1)^{\sigma_{\beta_k}^0 + \sigma_0^{\alpha_k}} = (-1)^{\sigma_{\beta_k}^{\alpha_k}}$ so, by the first fundamental formula, we have that (7.41) is equal to

$$(-1)^{\sum_{i=1}^k \sigma_{\beta_i}^{\alpha_i}} \langle e(f), \int_{0<t_1<\cdots<t_k<t} d\Xi_{\beta_1}^{\alpha_1}(t_1) \cdots d\Xi_{\beta_k}^{\alpha_k}(t_k) G(s)e(g) \rangle.$$

The proposition follows from the totality of the exponential vectors in $\mathcal{E}$.∎

We are now in a position to state and prove a proposition giving a formula for the adjoint of an iterated integral of the form (7.38).

Proposition 7.3 For arbitrary $t \geq 0$, arbitrary $n \geq 1$ and arbitrary α_i, β_i with $0 \leq \alpha_i, \beta_i \leq N$, $i = 1, \ldots, n$ we have

$$\left(\int_{0<t_1<\cdots<t_n<t} d\Xi_{\beta_1}^{\alpha_1}(t_1) \cdots d\Xi_{\beta_n}^{\alpha_n}(t_n) \right)^\dagger =$$

$$(-1)^{\sum_{1 \leq i < j \leq n} \sigma_{\beta_i}^{\alpha_i} \sigma_{\beta_j}^{\alpha_j}} \int_{0<t_1<\cdots<t_n<t} d\Xi_{\alpha_1}^{\beta_1}(t_1) \cdots d\Xi_{\alpha_n}^{\beta_n}(t_n).$$

$$(7.42)$$

Proof. We establish the result by means of induction and techniques that are standard in quantum stochastic calculus. If $n = 1$ then we have by (4.17) and lemma 4.6 that

$$\left(\int_{0<t_1<t} d\Xi_{\beta_1}^{\alpha_1}(t_1) \right)^\dagger = \int_{0<t_1<t} d\Xi_{\alpha_1}^{\beta_1}(t_1)$$

so the proposition holds in this case as the summation clause $1 \leq i < j \leq 1$ is not satisfiable. Thus we have the base for our induction.

Suppose that the result holds for all $n < k$ with k some integer. Now take $n = k$. Given arbitrary $t \geq 0$ and arbitrary α_i, β_i with $0 \leq \alpha_i, \beta_i \leq N$, $i = 1, \ldots, k$ we have from the string of equalities (7.37)

$$\left(\int_{0<t_1<\cdots<t_k<t} d\Xi_{\alpha_1}^{\beta_1}(t_1) \cdots d\Xi_{\alpha_k}^{\beta_k}(t_k) \right)^\dagger$$

$$(7.43)$$

$$= \int_0^t G^{\sigma_{\beta_k}^{\alpha_k}}(t_k) \left(\int_{0<t_1<\cdots<t_k} d\Xi_{\alpha_1}^{\beta_1}(t_1) \cdots d\Xi_{\alpha_{k-1}}^{\beta_{k-1}}(t_{k-1}) \right)^\dagger d\Lambda_{\alpha_k}^{\beta_k}(t_k).$$

Applying the inductive hypothesis, we have that the right-hand side of (7.43) is equal to

$$(-1)^{\sum_{1\leq i<j\leq k-1}\sigma_{\beta_i}^{\alpha_i}\sigma_{\beta_j}^{\alpha_j}}\int_0^t G^{\sigma_{\beta_k}^{\alpha_k}}(t_k)\int_{0<t_1<\cdots<t_{k-1}<t_k}d\Xi_{\alpha_1}^{\beta_1}(t_1)\cdots d\Xi_{\alpha_{k-1}}^{\beta_{k-1}}(t_{k-1})\,d\Lambda_{\alpha_k}^{\beta_k}(t_k).$$

$$(7.44)$$

By proposition 7.2, we have that (7.44) is equal to

$$\left((-1)^{\sum_{1\leq i<j\leq k-1}\sigma_{\beta_i}^{\alpha_i}\sigma_{\beta_j}^{\alpha_j}+\sigma_{\beta_k}^{\alpha_k}\sum_{i=1}^{k-1}\sigma_{\beta_i}^{\alpha_i}}\right)$$

$$\int_0^t\int_{0<t_1<\cdots<t_{k-1}<t_k}d\Xi_{\alpha_1}^{\beta_1}(t_1)\cdots d\Xi_{\alpha_{k-1}}^{\beta_{k-1}}(t_{k-1})G^{\sigma_{\beta_k}^{\alpha_k}}(t_k)\,d\Lambda_{\alpha_k}^{\beta_k}(t_k).$$

$$(7.45)$$

Modifying the clause $1\leq i<j\leq k-1$ to include the new grading factor and using the definition of $d\Xi_{\alpha_k}^{\beta_k}$ given by (4.16) we may re-write (7.45) as

$$(-1)^{\sum_{1\leq i<j\leq k}\sigma_{\beta_i}^{\alpha_i}\sigma_{\beta_j}^{\alpha_j}}\int_{0<t_1<\cdots<t_k<t}d\Xi_{\alpha_1}^{\beta_1}(t_1)\cdots d\Xi_{\alpha_k}^{\beta_k}(t_k)$$

and so the proposition holds.∎

7.5 The Map I

Let the Ito superalgebra $\mathcal{I}$ of Chapter 6 now be fixed as the space $\{d\Xi_A : A \in M_0(N,r)\}$ of $\mathbf{Z}_2$-graded quantum stochastic differentials under the quantum stochastic Ito multiplication described by (4.40) and (4.41). We might equally well fix $\mathcal{I}$ to be a sub-†-superalgebra of this space. By defining a map on the space $\mathcal{T}(\mathcal{I})$ of Chapter 6 we will construct a useful tool for dealing with iterated quantum stochastic integrals.

For each integer $n \geq 0$, define a map I^n on the space $\mathcal{I} \otimes \cdots \otimes \mathcal{I}$ (n copies of $\mathcal{I}$) by linear extension of the following rule for product tensors $d\Xi_1 \otimes \cdots \otimes d\Xi_n$:

$$I^n(d\Xi_1 \otimes \cdots \otimes d\Xi_n) = \int_{0<t_1<\cdots<t_n<t}d\Xi_1(t_1)\cdots d\Xi_n(t_n).$$

The map I^0 on $\mathbf{C}$ is defined by

$$I^0 : \mathbf{C} \ni z \mapsto z\,Id.$$

Recalling that the space $\mathcal{T}(\mathcal{I})$ is the direct sum

$$\mathbf{C} \oplus \mathcal{I} \oplus (\mathcal{I} \otimes \mathcal{I}) \oplus (\mathcal{I} \otimes \mathcal{I} \otimes \mathcal{I}) \oplus \cdots$$

we may amalgamate the I^n, $n \geq 0$ to form a map I on $\mathcal{T}(\mathcal{I})$ called the *integrator map*. We denote the image $\{I(a) : a \in \mathcal{T}(\mathcal{I})\}$ of I by $\mathcal{P}$.

An element of $\mathcal{P}$ is a quantum stochastic process formed from a complex sum of iterated quantum stochastic integrals. We denote the value of $I(a)$ at a time $t \geq 0$ by $I(a)_t$. Note that a comparison of (7.42) with (6.2) gives that $I(a)_t{}^\dagger = I(a^\dagger)_t$ for any $a \in T(\mathcal{I})$ and any time $t \geq 0$.

In general, an operator of the form $I(a)_t$ is unbounded and will not leave the exponential domain $\mathcal{E}$ invariant. Hence it is not possible to define a multiplication in $\mathcal{P}$ in the sense of composition of operators. However, in the next three sections we will develop a 'weak' but entirely rigorous and well-defined multiplication in $\mathcal{P}$.

7.6 Fundamental Property of $\star$

In this section we show that the map I is, in a certain sense, a unital $\dagger$-superalgebra morphism from $T(\mathcal{I})$ to $\mathcal{P}$.

Theorem 7.4 For arbitrary $t \geq 0$, arbitrary $\psi, \rho \in \mathcal{E}$ and arbitrary $a, b \in T(\mathcal{I})$ we have

$$\langle I(a)_t{}^\dagger \psi, I(b)_t \rho \rangle = \langle \psi, I(a \star b)_t \rho \rangle. \tag{7.46}$$

Proof. The proof is largely analogous to the proof of theorem 5.2 which is the ungraded version of this result. By linearity it suffices to prove the result for cases where

$$a = (0, \ldots, 0, a^1 \otimes \cdots \otimes a^m, 0, \cdots), \quad b = (0, \ldots, 0, b^1 \otimes \cdots \otimes b^n, 0, \cdots)$$

with $m, n \geq 0$ and each a^i, b^i a basis element of $\mathcal{I}$. Thus we may write $a^i = d\Xi^{\alpha_i}_{\beta_i}$, $b^j = d\Xi^{\gamma_j}_{\delta_j}$ for $1 \leq i \leq m$ and $1 \leq j \leq n$. Linearity also allows us to assume that ψ, ρ are exponential vectors of the Fock space $\Gamma(L^2(\mathbf{R}_+; \mathbf{C}^N))$ so we set $\psi = e(f)$ and $\rho = e(g)$ for some arbitrary $f, g \in L^2(\mathbf{R}_+; \mathbf{C}^N)$. By a slight abuse of notation we declare

$$a = a^1 \otimes \cdots \otimes a^m, \quad b = b^1 \otimes \cdots \otimes b^n.$$

If m or $n = 0$ then (7.46) is obvious so we may assume $m, n \geq 1$ and define

$$\dot{a} = a^1 \otimes \cdots \otimes a^{m-1}, \quad \dot{b} = b^1 \otimes \cdots \otimes b^{n-1}.$$

If $m = 1$ then we take $\dot{a} = 1 \in \mathbf{C}$ and, likewise, we take $\dot{b} = 1$ when $n = 1$.

We now state and prove a lemma.

Lemma 7.5 Using the notation already given

$$a \star b = (a \star \dot{b}) \otimes b^n + (-1)^{\sum_{i=1}^n \sigma(a^m)\sigma(b^i)} (\dot{a} \star b) \otimes a^m$$
$$+ (-1)^{\sum_{i=1}^{n-1} \sigma(a^m)\sigma(b^i)} (\dot{a} \star \dot{b}) \otimes a^m b^n.$$

Proof. (of lemma) Consider an arbitrary component of $a \star b$:

$$(a \star b)_k = \sum_{A \cup B = M_k} a_{|A|}{}^A \bullet^B b_{|B|} = \sum_{\substack{A \cup B = M_k \\ |A| = m \; |B| = n}} a^A \bullet^B b$$

$$= \sum_{\substack{A \cup D = M_{k-1} \\ |A| = m \; |D| = n-1}} \left(a^A \bullet^D \dot{b} \right) \otimes b^n$$

$$+ (-1)^{\sum_{i=1}^{n} \sigma(a^m)\sigma(b^i)} \sum_{\substack{C \cup B = M_{k-1} \\ |C| = m-1 \; |B| = n}} \left(\dot{a}^C \bullet^B b \right) \otimes a^m$$

$$+ (-1)^{\sum_{i=1}^{n-1} \sigma(a^m)\sigma(b^i)} \sum_{\substack{C \cup D = M_{k-1} \\ |C| = m-1 \; |D| = n-1}} \left(\dot{a}^C \bullet^D \dot{b} \right) \otimes a^m b^n$$

$$= \left((a \star \dot{b}) \otimes b^n + (-1)^{\sum_{i=1}^{n} \sigma(a^m)\sigma(b^i)} (\dot{a} \star b) \otimes a^m \right.$$

$$\left. + (-1)^{\sum_{i=1}^{n-1} \sigma(a^m)\sigma(b^i)} \left(\dot{a} \star \dot{b} \right) \otimes a^m b^n \right)_k$$

and so the the lemma is proved.

We now continue with the main part of the proof. Consider the case $m = n = 1$ so that $a = a^1 = d\Xi^{\alpha_1}_{\beta_1}$, $b = b^1 = d\Xi^{\gamma_1}_{\delta_1}$. We have already established that $I(a)^\dagger = I(a^\dagger)$ so, at an arbitrary time $t \geq 0$, we have that $\langle I(a)^\dagger_t e(f), I(b)_t e(g) \rangle$ is equal to

$$\langle \int_0^t d\Xi^{\beta_1}_{\alpha_1}(t)e(f), \int_0^t d\Xi^{\gamma_1}_{\delta_1}(t)e(g) \rangle. \tag{7.47}$$

Applying the particular form of the second fundamental formula given by (7.9) to this expression yields

$$\int_0^t f_{\delta_1}(s)g^{\gamma_1}(s)\langle \int_0^s d\Xi^{\beta_1}_{\alpha_1}(u)e(f), G^{\sigma^{\gamma_1}_{\delta_1}}(s)e(g) \rangle \, ds$$

$$+ \int_0^t f_{\beta_1}(s)g^{\alpha_1}(s)\langle G^{\sigma^{\beta_1}_{\alpha_1}}(s)e(f), \int_0^s d\Xi^{\gamma_1}_{\delta_1}(u)e(g) \rangle \, ds$$

$$+ \hat{\delta}^{\alpha_1}_{\delta_1} \int_0^t f_{\beta_1}(s)g^{\gamma_1}(s)\langle G^{\sigma^{\beta_1}_{\alpha_1}}(s)e(f), G^{\sigma^{\gamma_1}_{\delta_1}}(s)e(g) \rangle \, ds. \tag{7.48}$$

The self-adjointness of the operators $G(s)$ and Id allows the $G^{\sigma^{\beta_1}_{\alpha_1}}(s)$ in the third term of (7.48) to be moved to the right-hand side of the inner product. This yields $G^{\sigma^{\alpha_1}_{\beta_1} + \sigma^{\gamma_1}_{\delta_1}}(s)$. If $\hat{\delta}^{\alpha_1}_{\delta_1} \neq 0$ then $\alpha_1 = \delta_1$ so that $\sigma^{\alpha_1}_{\beta_1} + \sigma^{\gamma_1}_{\delta_1} = \sigma^{\gamma_1}_{\beta_1} \pmod 2$ by corollary 4.2. Using this fact and an application of the first fundamental formula (7.5) to the first two terms of (7.48) yields

$$\int_0^t f_{\delta_1}(s)g^{\gamma_1}(s)\int_0^s f_{\beta_1}(u)(-1)^{\sigma_{\delta_1}^{\gamma_1}\sigma_0^{\alpha_1}}g^{\alpha_1}(u)\langle G^{\sigma_{\alpha_1}^{\beta_1}}(u)e(f), G^{\sigma_{\delta_1}^{\gamma_1}}(s)e(g)\rangle\,du\,ds$$

$$+\int_0^t f_{\beta_1}(s)g^{\alpha_1}(s)\int_0^s (-1)^{\sigma_{\beta_1}^{\alpha_1}\sigma_{\delta_1}^0}f_{\delta_1}(u)g^{\gamma_1}(u)\langle G^{\sigma_{\alpha_1}^{\beta_1}}(s)e(f), G^{\sigma_{\delta_1}^{\gamma_1}}(u)e(g)\rangle\,du\,ds$$

$$+\hat{\delta}_{\delta_1}^{\alpha_1}\int_0^t f_{\beta_1}(s)g^{\gamma_1}(s)\langle e(f), G^{\sigma_{\beta_1}^{\gamma_1}}(s)e(g)\rangle\,ds. \tag{7.49}$$

The fact that $G(u)$, $G(s)$ and Id are all self-adjoint allows $G^{\sigma_{\alpha_1}^{\beta_1}}(u)$ and $G^{\sigma_{\alpha_1}^{\beta_1}}(s)$ to be moved to the right-hand side of their respective inner products. Lemma 4.3 gives us

$$G^{\sigma_{\alpha_1}^{\beta_1}}(s)G^{\sigma_{\delta_1}^{\gamma_1}}(u) = G^{\sigma_{\delta_1}^{\gamma_1}}(u)G^{\sigma_{\alpha_1}^{\beta_1}}(s)$$

so that (7.49) may be re-written as

$$(-1)^{\sigma_{\delta_1}^{\gamma_1}\sigma_0^{\alpha_1}}\int_0^t f_{\delta_1}(s)g^{\gamma_1}(s)\int_0^s f_{\beta_1}(u)g^{\alpha_1}(u)\langle e(f), G^{\sigma_{\beta_1}^{\alpha_1}}(u)G^{\sigma_{\delta_1}^{\gamma_1}}(s)e(g)\rangle\,du\,ds$$

$$+(-1)^{\sigma_{\beta_1}^{\alpha_1}\sigma_{\delta_1}^0}\int_0^t f_{\beta_1}(s)g^{\alpha_1}(s)\int_0^s f_{\delta_1}(u)g^{\gamma_1}(u)\langle e(f), G^{\sigma_{\delta_1}^{\gamma_1}}(u)G^{\sigma_{\beta_1}^{\alpha_1}}(s)e(g)\rangle\,du\,ds$$

$$+\hat{\delta}_{\delta_1}^{\alpha_1}\int_0^t f_{\beta_1}(s)g^{\gamma_1}(s)\langle e(f), G^{\sigma_{\beta_1}^{\gamma_1}}(s)e(g)\rangle\,ds. \tag{7.50}$$

We now apply the form of the first fundamental formula given in (7.5) to each of the terms in (7.50). In the first term we see that the $G^{\sigma_{\delta_1}^{\gamma_1}}(s)$ preceding the $e(g)$ makes it necessary to introduce a factor of $(-1)^{\sigma_{\beta_1}^{\alpha_1}\sigma_0^{\alpha_1}}$. In the second term of (7.50) we see that the $G^{\sigma_{\beta_1}^{\alpha_1}}(s)$ preceding the $e(g)$ necessitates the introduction of a factor of $(-1)^{\sigma_{\beta_1}^{\alpha_1}\sigma_0^{\gamma_1}}$. There are no complications in the application of the first fundamental formula to the third term of (7.50). Thus we re-write (7.50) as

$$(-1)^{\sigma_{\delta_1}^{\gamma_1}\sigma_0^{\alpha_1}+\sigma_{\delta_1}^{\gamma_1}\sigma_0^{\alpha_1}}\int_0^t f_{\delta_1}(s)g^{\gamma_1}(s)\langle e(f), \int_0^s d\Xi_{\beta_1}^{\alpha_1}(u)G^{\sigma_{\delta_1}^{\gamma_1}}(s)e(g)\rangle\,ds$$

$$+(-1)^{\sigma_{\beta_1}^{\alpha_1}\sigma_{\delta_1}^0+\sigma_{\beta_1}^{\alpha_1}\sigma_0^{\gamma_1}}\int_0^t f_{\beta_1}(s)g^{\alpha_1}(s)\langle e(f), \int_0^s d\Xi_{\delta_1}^{\gamma_1}(u)G^{\sigma_{\beta_1}^{\alpha_1}}(s)e(g)\rangle\,ds$$

$$+\hat{\delta}_{\delta_1}^{\alpha_1}\langle e(f), \int_0^t d\Xi_{\beta_1}^{\gamma_1}(s)e(g)\rangle. \tag{7.51}$$

Clearly $(-1)^{\sigma_{\delta_1}^{\gamma_1}\sigma_0^{\alpha_1}+\sigma_{\delta_1}^{\gamma_1}\sigma_0^{\alpha_1}} = 1$. We also know by corollary 4.2 that $\sigma_{\delta_1}^0 + \sigma_0^{\gamma_1} = \sigma_{\delta_1}^{\gamma_1}$ (mod 2) so that $(-1)^{\sigma_{\beta_1}^{\alpha_1}\sigma_{\delta_1}^0+\sigma_{\beta_1}^{\alpha_1}\sigma_0^{\gamma_1}} = (-1)^{\sigma_{\beta_1}^{\alpha_1}\sigma_{\delta_1}^{\gamma_1}}$. These observations and an uncomplicated application of the first fundamental formula to the first two terms of (7.51) gives us that (7.51) is equal to

$$\langle e(f), \int_0^t \int_0^s d\Xi_{\beta_1}^{\alpha_1}(u)\, d\Xi_{\delta_1}^{\gamma_1}(s) e(g)\rangle$$

$$+(-1)^{\sigma_{\beta_1}^{\alpha_1}\sigma_{\delta_1}^{\gamma_1}} \langle e(f), \int_0^t \int_0^s d\Xi_{\delta_1}^{\gamma_1}(u) d\Xi_{\beta_1}^{\alpha_1}(s) e(g)\rangle \qquad (7.52)$$

$$+\hat{\delta}_{\delta_1}^{\alpha_1}\langle e(f), \int_0^t d\Xi_{\beta_1}^{\gamma_1}(s) e(g)\rangle.$$

By the $\mathbf{Z}_2$-graded quantum Ito's formula (4.40) we have that

$$\hat{\delta}_{\delta_1}^{\alpha_1} d\Xi_{\beta_1}^{\gamma_1} = d\Xi_{\beta_1}^{\alpha_1}.d\Xi_{\delta_1}^{\gamma_1} = a^1 b^1.$$

Thus, switching to the more algebraic notation, we re-write (7.52) as

$$\langle e(f), I(a^1 \otimes a^2)_t e(g)\rangle + (-1)^{\sigma(a^1)\sigma(b^1)}\langle e(f), I(b^1 \otimes a^1)_t e(g)\rangle$$
$$+ \langle e(f), I(a^1 b^1)_t e(g)\rangle. \qquad (7.53)$$

The definition of the $\star$ product given in Section 6.4 shows that

$$(a \star b)_0 = 0;$$
$$(a \star b)_1 = a^{\{1\}} \bullet^{\{1\}} b = a^1 b^1;$$
$$(a \star b)_2 = \sum_{A \cup B = \{1,2\}} a^A \bullet^B b$$
$$= a^{\{1\}} \bullet^{\{2\}} b + a^{\{2\}} \bullet^{\{1\}} b$$
$$= a^1 \otimes b^1 + (-1)^{\sigma(a^1)\sigma(b^1)} b^1 \otimes a^1;$$
$$(a \star b)_k = 0 \text{ for all } k \geq 3.$$

Thus (7.53) is equal to

$$\langle e(f), I(a \star b)_t e(g)\rangle.$$

We now have that the theorem holds in all cases where $n + m \leq 2$. This forms the base for our induction which is on the value of $m + n$. Now suppose the theorem holds for all m, n such that $m + n < k$ for some integer k. Take some a and b in $T(\mathcal{I})$ such that $m + n = k$. Before proceeding we present another lemma.

Lemma 7.6: For arbitrary $c = d\Xi_{\mu_1}^{\epsilon_1} \otimes \cdots \otimes d\Xi_{\mu_l}^{\epsilon_l}$ in $T(\mathcal{I})$ of non-zero degree we have that for arbitrary $t \geq 0$

$$I(c)_t^\dagger = (-1)^{\sum_{i=1}^{l-1} \sigma_{\mu_i}^{\epsilon_i}\sigma_{\mu_l}^{\epsilon_l}} \int_0^t I(\dot{c}^\dagger)_s \, d\Xi_{\epsilon_l}^{\mu_l}(s)$$

where $\dot{c} := d\Xi_{\mu_1}^{\epsilon_1} \otimes \cdots \otimes d\Xi_{\mu_{l-1}}^{\epsilon_{l-1}}$.

Proof of lemma: We may re-write $I(c)_t^\dagger$ as

$$\left(\int_0^t I(\dot{c})_s G^{\sigma_{\mu_l}^{\epsilon_l}}(s) \, d\Lambda_{\mu_l}^{\epsilon_l}\right)^\dagger$$

which, by standard theory, is equal to

$$\int_0^t G^{\sigma_{\mu_l}^{\epsilon_l}}(s)I(\dot{c})_s^\dagger \, d\Lambda_{\epsilon_l}^{\mu_l}(s). \tag{7.54}$$

By proposition 7.2 and the fact that for all e in $T(\mathcal{I})$ we have $I(e)^\dagger = I(e^\dagger)$ we may re-write (7.54) as

$$(-1)^{\sum_{i=1}^{l-1} \sigma_{\mu_l}^{\epsilon_l} \sigma_{\mu_i}^{\epsilon_i}} \int_0^t I(\dot{c}^\dagger)_s G^{\sigma_{\mu_l}^{\epsilon_l}}(s) \, d\Lambda_{\epsilon_l}^{\mu_l}(s)$$

which, by (4.16), is equal to

$$(-1)^{\sum_{i=1}^{l-1} \sigma_{\mu_l}^{\epsilon_l} \sigma_{\mu_i}^{\epsilon_i}} \int_0^t I(\dot{c}^\dagger)_s \, d\Xi_{\epsilon_l}^{\mu_l}(s)$$

as required.

We now continue with the main proof. Set $Y' := (-1)^{\sum_{i=1}^{m-1} \sigma(a^m)\sigma(a^i)}$ as a matter of typographical convenience. Using lemma 7.6, we may re-write $\langle I(a)_t{}^\dagger e(f), I(b)_t e(g)\rangle$ as

$$Y'\langle \int_0^t I(\dot{a}^\dagger)_s \, d\Xi_{\alpha_m}^{\beta_m}(s)e(f), \int_0^t I(\dot{b})_s \, d\Xi_{\delta_n}^{\gamma_n}(s)e(g)\rangle. \tag{7.55}$$

By the version of the second fundamental formula given in (7.12) we can see that (7.55) is equal to

$$\int_0^t f_{\delta_n}(s)g^{\gamma_n}(s)\langle I(a^\dagger)_s e(f), I(\dot{b})_s G^{\sigma_{\delta_n}^{\gamma_n}}(s)e(g)\rangle \, ds$$

$$+Y' \int_0^t f_{\beta_m}(s)g^{\alpha_m}(s)\langle I(\dot{a}^\dagger)_s G^{\sigma_{\alpha_m}^{\beta_m}}(s)e(f), I(b)_s e(g)\rangle \, ds \tag{7.56}$$

$$+Y'\hat{\delta}_{\delta_n}^{\alpha_m} \int_0^t f_{\beta_m}(s)g^{\gamma_n}(s)\langle I(\dot{a}^\dagger)_s G^{\sigma_{\alpha_m}^{\beta_m}}(s)e(f), I(\dot{b})_s G^{\sigma_{\delta_n}^{\gamma_n}}(s)e(g)\rangle \, ds.$$

Invoking the inductive hypothesis and using the self-adjointness of $G^{\sigma_{\beta_m}^{\alpha_m}}(s)$ we may re-write (7.56) as

$$\int_0^t f_{\delta_n}(s)g^{\gamma_n}(s)\langle e(f), I(a \star \dot{b})G^{\sigma_{\delta_n}^{\gamma_n}}(s)e(g)\rangle \, ds$$

$$+Y' \int_0^t f_{\beta_m}(s)g^{\alpha_m}(s)\langle e(f), G^{\sigma_{\beta_m}^{\alpha_m}}(s)I(\dot{a} \star b)_s e(g)\rangle \, ds \tag{7.57}$$

$$+Y'\hat{\delta}_{\delta_n}^{\alpha_m} \int_0^t f_{\beta_m}(s)g^{\gamma_n}(s)\langle e(f), G^{\sigma_{\beta_m}^{\alpha_m}}(s)I(\dot{a} \star \dot{b})G^{\sigma_{\delta_n}^{\gamma_n}}(s)e(g)\rangle \, ds.$$

Proposition 7.2 may be applied to the second and third terms of (7.57) if some care is taken. Each entry of each product tensor that comprises $\dot{a} \star b$

is of the form either a^i, b^j or $a^i b^j$ with each entry of $\dot a = a^1 \otimes \cdots \otimes a^{m-1}$ and $b = b^1 \otimes \cdots \otimes b^n$ appearing exactly once. The parity $\sigma(a^i b^j)$ of $a^i b^j$ is $\sigma(a^i) + \sigma(b^j)$. Hence the parity of each product tensor comprising $\dot a \star b$ is $\sum_{i=1}^{m-1} \sigma(a^i) + \sum_{j=1}^{n} \sigma(b^j)$. We may conclude from this that the parity of $\dot a \star b$ is $\sum_{i=1}^{m-1} \sigma(a^i) + \sum_{j=1}^{n} \sigma(b^j)$. It can be shown in a similar fashion that the parity of $\dot a \star \dot b$ is $\sum_{i=1}^{m-1} \sigma(a^i) + \sum_{j=1}^{n-1} \sigma(b^j)$. Thus, proposition 7.2 gives us that

$$G^{\sigma(a^m)}(s) I(\dot a \star b)_s = (-1)^{\sum_{i=1}^{m-1} \sigma(a^m)\sigma(a^i) + \sum_{j=1}^{n} \sigma(a^m)\sigma(b^j)} I(\dot a \star b)_s G(s) \tag{7.58}$$

and

$$G^{\sigma(a^m)}(s) I(\dot a \star \dot b)_s = (-1)^{\sum_{i=1}^{m-1} \sigma(a^m)\sigma(a^i) + \sum_{j=1}^{n-1} \sigma(a^m)\sigma(b^j)} I(\dot a \star \dot b)_s G(s). \tag{7.59}$$

Cancelling Y' with the power of -1 on the right of (7.58) for the second term of (7.57) and cancelling Y' with the power of -1 on the right of (7.59) for the third term of (7.57) we see that (7.57) may be re-written as

$$\int_0^t f_{\delta_n}(s) g^{\gamma_n}(s) \langle e(f), I(a \star \dot b)_s G^{\sigma_{\delta_n}^{\gamma_n}}(s) e(g) \rangle \, ds$$

$$+ (-1)^{\sum_{j=1}^{n} \sigma(a_m)\sigma(b_j)} \int_0^t f_{\beta_m}(s) g^{\alpha_m}(s) \langle e(f), I(\dot a \star b)_s G^{\sigma_{\beta_m}^{\alpha_m}}(s) e(g) \rangle \, ds$$

$$+ (-1)^{\sum_{j=1}^{n-1} \sigma(a_m)\sigma(b_j)} \hat\delta_{\delta_n}^{\alpha_m} \int_0^t f_{\beta_m}(s) g^{\gamma_n}(s) \langle e(f), I(\dot a \star \dot b)_s G^{\sigma_{\beta_m}^{\gamma_n}}(s) e(g) \rangle \, ds. \tag{7.60}$$

An application of the first fundamental formula as given in (7.5) to each term of (7.60) followed by an application of (4.40) to the third term of (7.60) gives us that (7.60) is equal to

$$\langle e(f), \int_0^t I(a \star \dot b)_s \, d\Xi_{\delta_n}^{\gamma_n}(s) e(g) \rangle$$

$$+ \langle e(f), (-1)^{\sum_{j=1}^{n} \sigma(a^m)\sigma(b^j)} \int_0^t I(\dot a \star b)_s \, d\Xi_{\beta_m}^{\alpha_m}(s) e(g) \rangle \tag{7.61}$$

$$+ \langle e(f), (-1)^{\sum_{j=1}^{n-1} \sigma(a^m)\sigma(b^j)} \int_0^t I(\dot a \star \dot b)_s \, d\Xi_{\beta_m}^{\alpha_m} . d\Xi_{\delta_n}^{\gamma_n}(s) e(g) \rangle.$$

A complete reversion to the algebraic notation re-writes (7.61) as

$$\langle e(f), I\Big((a \star \dot b) \otimes b^n + (-1)^{\sum_{j=1}^{n} \sigma(a^m)\sigma(b^j)} (\dot a \star b) \otimes a^m$$

$$+ (-1)^{\sum_{j=1}^{n-1} \sigma(a^m)\sigma(b^j)} (\dot a \star \dot b) \otimes a^m b^n \Big)_t e(g) \rangle$$

which, by lemma 7.5, is equal to

$$\langle e(f), I(a \star b)_t e(g) \rangle$$

and so the theorem holds. ∎

7.7 The Injectivity of I

In the next section we will introduce a rigorous product defined on the the space $\mathcal{P}$. The requirement that this product should be well-defined relies on the injectivity of the map I. Indeed, that I is injective is a result of intrinsic interest. In this section we prove this injectivity using a result from [L].

Proposition 7.7 Let $((a_\beta^\alpha))$ be a matrix of elements of $\mathcal{T}(\mathcal{I})$. If it is true that for each $t \geq 0$ we have $\int_0^t I(a_\beta^\alpha) \, d\Xi_\alpha^\beta = 0$ then it follows that for each α, β we have $I(a_\beta^\alpha) = 0$.

Proof. The proposition follows as a special case of corollary 1.3 in [L]. This corollary will be reviewed here. First it is necessary to introduce some notation. The set $\mathcal{D}_S$ described in [L] may be taken here to be the exponential domain $\mathcal{E}$ defined in Chapter 2. A process X is said to be a $\mathcal{D}_S$-process if:

      (a) $\mathrm{Dom} X(t) \supset \mathcal{D}_S \quad \forall t \in \mathbf{R}_+$,

      (b) $t \mapsto X(t)k$ is Borel measurable $\forall k \in \mathcal{D}_S$.

A $\mathcal{D}_S$-process X is said to be *locally square integrable* if, for each $T \geq 0$ and each $k \in \mathcal{D}_S$, we have

$$\int_0^T \|X(t)k\|^2 \, dt < \infty.$$

Corollary 1.3 of [L] states that if $((F_\beta^\alpha))$ is a matrix of adapted, locally square integrable $\mathcal{D}_S$-processes and each operator $F_\beta^\alpha(t)$, $t \in \mathbf{R}_+$ is closable then

$$\int_0^\cdot F_\beta^\alpha \, d\Lambda_\alpha^\beta = 0 \text{ implies that } ((F_\beta^\alpha)) = 0.$$

In the case under consideration we have an integral

$$\int_0^\cdot I(a_\beta^\alpha) \, d\Xi_\alpha^\beta = \sum_{\alpha,\beta=0}^N \int_0^\cdot I(a_\beta^\alpha) G^{\sigma_\alpha^\beta} \, d\Lambda_\alpha^\beta.$$

It was seen in Chapter 2 that quantum stochastic integrals are adapted processes so we have that each $I(a_\beta^\alpha)$ is an adapted process. It follows that for each α, β with $0 \leq \alpha, \beta \leq N$ the process $I(a_\beta^\alpha) G^{\sigma_\alpha^\beta} = \left(I(a_\beta^\alpha)_t G^{\sigma_\alpha^\beta}(t) \right)_{t \geq 0}$ is adapted. For each α, β and each $t \geq 0$, the operator $I(a_\beta^\alpha)_t G^{\sigma_\alpha^\beta}(t)$ has an adjoint $G^{\sigma^\beta_\alpha}(t)I(a^\alpha_\beta{}^\dagger)_t$. It follows from this that each $I(a^\alpha_\beta)_t G^{\sigma^\beta_\alpha}(t)$ is closable. For each α, β, the process $I(a^\alpha_\beta)$ is defined on $\mathcal{E} = \mathcal{D}_S$ and, as G and Id are both defined on the whole Fock space $\Gamma(L^2(\mathbf{R}_+; \mathbf{C}^N))$, the process $I(a^\alpha_\beta)G^{\sigma^\beta_\alpha}$ is certainly defined on the whole of $\mathcal{E}$. Thus condition (a) of the $\mathcal{D}_S$-process property is satisfied. It was stated in proposition 2.16 that quantum stochastic integrals are are time-continuous and are therefore Borel measurable in the sense of (b). This continuity also provides the required square-integrability property. Thus we have that each of the $I(a^\alpha_\beta)$ is a $\mathcal{D}_S$-process. All the conditions on the $I(a^\alpha_\beta)$ now being satisfied, we may conclude that $\int_0^\cdot I(a^\alpha_\beta) d\Xi^\beta_\alpha = 0$ implies that for all $0 \leq \alpha, \beta \leq N$ we have

$$I(a^\alpha_\beta)G^{\sigma^\beta_\alpha} = 0. \tag{7.62}$$

From (7.62) we may conclude directly that all $I(a^\alpha_\beta)$ with $\sigma^\alpha_\beta = 0$ are zero. The remaining terms have $\sigma^\alpha_\beta = 1$ and, as the map G leaves the exponential domain $\mathcal{E}$ invariant, we may multiply each side of (7.62) by G yielding $I(a^\alpha_\beta) = 0$ as required. Having established that all the terms are zero, the result follows.∎

Proposition 7.8 The map $I : \mathcal{T}(\mathcal{I}) \to \mathcal{P}$ is injective.

Proof. The map I is linear so it suffices to show that, given an arbitrary $a \in \mathcal{T}(\mathcal{I})$, $I(a) = 0$ implies $a = 0$.

Suppose a is of the form $(a_0, a_1, a_2, \ldots)$ where $a_0 \neq 0$. Then $I(a)_0 = a_0 Id \neq 0$ so that $I(a) \neq 0$. It remains to show that the proposition holds in the case $a_0 = 0$.

We prove the proposition for elements of the form $(0, a_1, a_2, \ldots)$ by induction on the order of the highest-order non-zero term. Note that, by the definition of $\mathcal{T}(\mathcal{I})$, this is always finite.

Suppose that $a = (0, a_1, 0, \ldots)$. Then $I(a) = \int I(a^\beta_\alpha) d\Xi^\alpha_\beta$ where each $a^\beta_\alpha \in \mathbf{C}$. By proposition 7.7, if $I(a) = 0$ we have that $I(a^\beta_\alpha) = 0$ for each a^β_α. By definition, for $z \in \mathbf{C}$ we have $I(z) = zId$ so if $I(z) = 0$ it is certainly true that $z = 0$. Therefore we must have that each $a^\beta_\alpha = 0$. By construction, $a = a^\alpha_\beta d\Xi^\beta_\alpha$ so we have $a = 0$ and the base for the induction is established.

Suppose now that the proposition holds for all $a \in \mathcal{T}(\mathcal{I})$ that are of the form $(0, a_1, \ldots, a_n, 0, \ldots)$ where $n < k$ for some integer k. Now take $a = (0, a_1, \ldots, a_k, 0, 0, \ldots)$. Then $I(a) = \int_0^\cdot I(a^\alpha_\beta) d\Xi^\beta_\alpha$ for some $a^\alpha_\beta \in \mathcal{T}(\mathcal{I})$. By proposition 7.7 we have $I(a^\alpha_\beta) = 0$ for each α, β. It is clear that the degree of each a^α_β is smaller than k so that the inductive hypothesis holds and we have that each $a^\alpha_\beta = 0$. By construction, $a = a^\alpha_\beta \otimes d\Xi^\beta_\alpha$ so it follows that $a = 0$ as required. The proposition now follows by induction.∎

7.8 The $\odot$ Product

Recall that $\mathcal{P}$ is defined to be $\{I(a) : a \in T(\mathcal{I})\}$, the space of all complex linear combinations of iterated $\mathbf{Z}_2$-graded quantum stochastic integrals. The results of the previous two sections allow us to define a rigorous product in $\mathcal{P}$. If A is an element of $\mathcal{P}$ so that $A = I(a)$ for some $a \in T(\mathcal{I})$ and B is an element of $\mathcal{P}$ so that $B = I(b)$ for some $b \in T(\mathcal{I})$ then we define the product $A \odot B$ by

$$A \odot B = I(a \star b). \tag{7.63}$$

This multiplication is well-defined by the injectivity of I proved in the previous section.

The importance of the product $\odot$ lies in the fact that it provides a rigorous interpretation of the formal products of unbounded operators given earlier in this chapter. A formal product XY of processes is intended to be interpreted in the 'weak' sense as

$$\langle X(t)^\dagger e(f), Y(t)e(g) \rangle$$

where t runs over all of $\mathbf{R}_+$ and $e(f), e(g)$ are exponential vectors with f, g being arbitrary elements of $L^2(\mathbf{R}_+; \mathbf{C}^N)$. With A and B as above, we have for an arbitrary time $t \geq 0$ that

$$\langle e(f), (A \odot B)_t e(g) \rangle = \langle e(f), I(a \star b)_t e(g) \rangle. \tag{7.64}$$

Theorem 7.4 shows that the right-hand side of (7.64) is equal to

$$\langle I(a)_t{}^\dagger e(f), I(b)_t e(g) \rangle$$

which is equal to

$$\langle A(t)^\dagger e(f), B(t)e(g) \rangle.$$

Thus we see that the product $A \odot B$ yields a well-defined process in $\mathcal{P}$ which is defined on all of the exponential domain $\mathcal{E}$. This product corresponds to the formal product AB requiring interpretation in terms of the inner product and adjoints.

Proposition 7.9 The pair $(\mathcal{P}, \odot)$ forms a unital complex associative †-superalgebra.

Proof. The process $Id = I((1, 0, 0, \ldots))$ is an element of $\mathcal{P}$ and it is clear from the definition given in (7.63) that for all $A \in \mathcal{P}$ we have $A \odot Id = Id \odot A = A$. Thus $(\mathcal{P}, \odot)$ is unital. It is clear from corollary 6.3 that the space $\mathcal{P}$ is closed under $\odot$. The associativity of $\star$ given by theorem 6.2 provides the required associativity of $\odot$. We also have from theorem 6.2 that

$$(A \odot B)^\dagger = I(a \star b)^\dagger = I((a \star b)^\dagger) = I(b^\dagger \star a^\dagger) = B^\dagger \odot A^\dagger$$

so that the antimultiplicity of the involution property holds. That the other properties required of the involution hold is obvious. It follows from proposition 7.1 and theorem 6.2 that $\sigma(A \odot B) = \sigma(A) + \sigma(B)$ so we have that $(\mathcal{P}, \odot)$ is indeed a superalgebra.∎

8 Chaotic Expansions

8.1 Preliminaries

Let $\mathcal{I}$ be a †-superalgebra of $\mathbf{Z}_2$-graded multidimensional quantum stochastic differentials. Hence $\mathcal{I}$ is an Ito superalgebra as defined in Section 6.1. Consider the corresponding Lie †-superalgebra $\mathcal{I}_{SLie}$ realised by equipping the associative superalgebra $\mathcal{I}$ with the supercommutator bracket described in Section 3.3. This bracket is defined by bilinear extension of the following rule for homogeneous $d\Xi_1, d\Xi_2$ in $\mathcal{I}_{SLie}$:

$$\{d\Xi_1, d\Xi_2\}_{\mathcal{I}_{SLie}} = d\Xi_1.d\Xi_2 - (-1)^{\sigma(d\Xi_1)\sigma(d\Xi_2)}d\Xi_2.d\Xi_1.$$

Let $\mathcal{L}$ be an abstract Lie †-superalgebra that is isomorphic to $\mathcal{I}_{SLie}$ by means of a map $L \mapsto d\Xi_L$. Let $\mathcal{U}$ be the universal enveloping superalgebra of $\mathcal{L}$. The notion of a universal enveloping superalgebra was discussed in Section 3.4. The superalgebra $\mathcal{U}$ is a complex unital associative †-superalgebra equipped with an injection $\imath: \mathcal{L} \to \mathcal{U}_{SLie}$. The injection $\imath$ has the *universal property*. Given a Lie †-superalgebra morphism $\phi: \mathcal{L} \to \mathcal{A}_{SLie}$ where $\mathcal{A}$ is an arbitrary complex unital associative superalgebra there will exist a unique complex unital associative superalgebra morphism $\tilde{\phi}: \mathcal{U} \to \mathcal{A}$ such that $\tilde{\phi} \circ \imath = \phi$. The map $\tilde{\phi}$ is said to *extend* ϕ. As described in Section 3.4, the map $\imath$ is used to identify $\mathcal{L}$ as a subsuperalgebra of $\mathcal{U}_{SLie}$.

Recall that for a map θ between two Lie †-superalgebras $\mathcal{N}$ and $\mathcal{N}'$ to be a Lie †-superalgebra morphism we require that it possesses three properties. Firstly, we require that for all $n \in \mathcal{N}$ we have that $\theta(n^\dagger) = \theta(n)^\dagger$. Secondly we require that θ is parity preserving, that is to say for each homogeneous n in $\mathcal{N}$ we require that $\sigma(n) = \sigma(\theta(n))$. Thirdly we require that for all n_1, n_2 in $\mathcal{N}$ we have

$$\theta(\{n_1, n_2\}_{\mathcal{N}}) = \{\theta(n_1), \theta(n_2)\}_{\mathcal{N}'}.$$

Proposition 7.9 states that $(\mathcal{P}, \odot)$ is a unital complex associative †-superalgebra. This means that the Lie †-superalgebra $\mathcal{P}_{SLie}$ may be constructed as described in Section 3.3. In particular, the superbracket of $\mathcal{P}_{SLie}$ will be defined by bilinear extension of the following rule for homogeneous elements A, B of $\mathcal{P}$:

$$\{A, B\}_{\mathcal{P}_{SLie}} = A \odot B - (-1)^{\sigma(A)\sigma(B)}B \odot A.$$

We now state and prove a theorem.

Theorem 8.1 The map $\phi\colon \mathcal{L} \to \mathcal{P}_{SLie}$ defined by $L \mapsto \Xi_L$ is a linear Lie †-superalgebra morphism.

Proof. It suffices to prove the result for $L, M \in \mathcal{L}$ of definite parity. For typographical convenience we set $Y = (-1)^{\sigma(L)\sigma(M)}$. The fact that the map $L \mapsto d\Xi_L$ is assumed to be grade preserving gives us that for all $N \in \mathcal{L}$ we have $\sigma(d\Xi_N) = \sigma(N)$. It follows by (4.17) and proposition 7.1 that $\sigma(\Xi_N) = \sigma(N)$ for all $N \in \mathcal{L}$. Thus we have

$$
\begin{aligned}
\{\phi(L), \phi(M)\}_{\mathcal{P}_{SLie}} &= \Xi_L \odot \Xi_M - Y \Xi_M \odot \Xi_L \\
&= I(d\Xi_L \star d\Xi_M) - Y I(d\Xi_M \star d\Xi_L) \\
&= I(0, d\Xi_L.d\Xi_M, d\Xi_L \otimes d\Xi_M + Y d\Xi_M \otimes d\Xi_L, 0, \ldots) \\
&\quad - Y I(0, d\Xi_M.d\Xi_L, d\Xi_M \otimes d\Xi_L + Y d\Xi_L \otimes d\Xi_M, 0, \ldots) \\
&= I(0, d\Xi_L.d\Xi_M - Y d\Xi_M.d\Xi_L, 0, \ldots) \\
&= I(0, d\Xi_{\{L,M\}_{\mathcal{L}}}, 0, \ldots) \\
&= \Xi_{\{L,M\}_{\mathcal{L}}} \\
&= \phi(\{L, M\}_{\mathcal{L}}).
\end{aligned}
$$

Therefore ϕ is a Lie superalgebra morphism. We have assumed that $L \mapsto d\Xi_L$ is a †-morphism and it was established in corollary 4.7. that $\int_0^t (d\Xi_L)^\dagger = \left(\int_0^t d\Xi_L\right)^\dagger$. It follows that $\phi(L^\dagger) = \phi(L)^\dagger$ for all $L \in \mathcal{L}$ and therefore ϕ is a †-morphism. It is obvious that ϕ is a linear map and so the result follows.∎

Corollary 8.2 There exists a unique complex unital †-superalgebra morphism J from $\mathcal{U}$ to $\mathcal{P}$ such that $J(L) = \Xi_L$ for each $L \in \mathcal{L}$.

Proof. The universal property of $\mathcal{U}$ guarantees the existence of a map that extends the Lie †-superalgebra morphism $\phi\colon \mathcal{L} \to \mathcal{P}_{SLie}$ described in theorem 8.1. Such an extension is the required map J.∎

8.2 The Co-Unit

The complex numbers $\mathbf{C}$ may be considered as a unital associative †-superalgebra with even space $\mathbf{C}$ and odd space $\{0\}$. The superbracket of $\mathbf{C}_{SLie}$ will map each pair of elements of $\mathbf{C}_{SLie}$ to zero as the multiplication in $\mathbf{C}$ is commutative. The involution † on $\mathbf{C}$ must be complex conjugation so that $z^\dagger = \bar{z}$.

Consider the zero map $\mathcal{L} \to \mathbf{C}_{SLie}$ that sends each element of $\mathcal{L}$ to 0. This is certainly a Lie †-superalgebra morphism mapping to a unital Lie †-superalgebra. It follows from the universal property of $\mathcal{U}$ that there exists a unique unital complex †-superalgebra morphism extending this zero map to all of $\mathcal{U}$. This map is denoted η and is known as the *co-unit*. The kernel of η is denoted $\mathcal{K}$ and is equal to the ideal of $\mathcal{U}$ generated by $\mathcal{L}$, that is to say, $\mathcal{K} = \mathcal{ULU}$. Note that η is not the zero map on $\mathcal{U}$. As detailed in Chapter 3, the extension $\tilde{\psi}$ of a Lie †-superalgebra morphism $\psi \colon \mathcal{Q} \to \mathcal{R}_{SLie}$ has $\tilde{\psi}(1_{\mathcal{U}_{\mathcal{Q}}}) = 1_{\mathcal{R}}$ where $1_{\mathcal{U}_{\mathcal{Q}}}$ is the unit of the universal enveloping superalgebra $\mathcal{U}_{\mathcal{Q}}$ associated with the Lie †-superalgebra $\mathcal{Q}$ and $1_{\mathcal{R}}$ is the unit of the †-superalgebra $\mathcal{R}$. Applying this general principle to the case in hand we see that $\eta(1_{\mathcal{U}}) = 1 \in \mathbf{C}$ so that any element of $\mathcal{U}$ not contained in $\mathcal{K} = \mathcal{ULU}$ has a non-zero image under η.

Proposition 8.3 Let $\mathcal{B}$ be a complex associative †-superalgebra not necessarily possessing a unit. Let $\phi_1 \colon \mathcal{L} \to \mathcal{B}_{SLie}$ be a Lie †-superalgebra morphism. Then ϕ_1 may be extended uniquely to a †-superalgebra morphism $\tilde{\phi}_1 \colon \mathcal{K} \to \mathcal{B}$.

Proof. Denote by $\mathcal{A}$ the space $\mathcal{B} \times \mathbf{C}$ equipped with multiplication

$$(b_1, z_1)(b_2, z_2) = (b_1 b_2 + z_1 b_2 + b_1 z_2, z_1 z_2). \tag{8.1}$$

We define an involution on $\mathcal{A}$ by prescribing that, for arbitrary $b \in \mathcal{B}$ and arbitrary $z \in \mathbf{C}$,

$$(b, z)^\dagger = (b^\dagger, \bar{z}). \tag{8.2}$$

This is certainly an involution as for arbitrary $b_1, b_2 \in \mathcal{B}$ and arbitrary $z_1, z_2 \in \mathbf{C}$ we have

$$\begin{aligned}
((b_1, z_1)(b_2, z_2))^\dagger &= (b_1 b_2 + z_1 b_2 + b_1 z_2, z_1 z_2)^\dagger \\
&= (b_2{}^\dagger b_1{}^\dagger + b_2{}^\dagger \bar{z}_1 + \bar{z}_2 b_1{}^\dagger, \bar{z}_2 \bar{z}_1) \\
&= (b_2{}^\dagger, \bar{z}_2)(b_1{}^\dagger, \bar{z}_1) \\
&= (b_2, z_2)^\dagger (b_1, z_1)^\dagger.
\end{aligned}$$

The element $(0, 1) \in \mathcal{A}$ possesses the property that, given arbitrary $(b, z) \in \mathcal{A}$, we have

$$(0, 1)(b, z) = (0b + 1b + 0z, 1z) = (b, z);$$
$$(b, z)(0, 1) = (b0 + z0 + b1, z1) = (b, z).$$

From this pair of results we may conclude that $(0, 1)$ is a unit for $\mathcal{A}$. We now show that $\mathcal{A}$ is associative. Given arbitrary $(b_1, z_1), (b_2, z_2), (b_3, z_3) \in \mathcal{A}$ we have

$$\begin{aligned}
&((b_1, z_1)(b_2, z_2))\,(b_3, z_3) \\
&= (b_1 b_2 + z_1 b_2 + b_1 z_2, z_1 z_2)(b_3, z_3) \\
&= (b_1 b_2 b_3 + z_1 b_2 b_3 + b_1 z_2 b_3 + z_1 z_2 b_3 + b_1 b_2 z_3 + z_1 b_2 z_3 + b_1 z_2 z_3, z_1 z_2 z_3) \\
&= (b_1 b_2 b_3 + b_1 z_2 b_3 + b_1 b_2 z_3 + z_1 b_2 b_3 + z_1 z_2 b_3 + z_1 b_2 z_3 + b_1 z_2 z_3, z_1 z_2 z_3) \\
&= ((b_1, z_1)\,((b_2 b_3 + z_2 b_3 + b_2 z_3, z_2 z_3)) \\
&= (b_1, z_1)\,((b_2, z_2)(b_3, z_3))
\end{aligned}$$

so that associativity for $\mathcal{A}$ holds.

Denote the even subspace of the superalgebra $\mathcal{B}$ by $\mathcal{B}_0$ and the odd subspace of $\mathcal{B}$ by $\mathcal{B}_1$. Define $\mathcal{A}_0$ to be the subspace $\mathcal{B}_0 \times \mathbf{C}$ of $\mathcal{A}$ and $\mathcal{A}_1$ to be the subspace $\mathcal{B}_1 \times \{0\}$ of $\mathcal{A}$. It is clear that $\mathcal{A} = \mathcal{A}_0 + \mathcal{A}_1$. Taking arbitrary $(b_0, z), (b_0', z')$ in $\mathcal{A}_0$ and arbitrary $(b_1, 0), (b_1', 0)$ in $\mathcal{A}_1$ we have

$$
\begin{aligned}
(b_0, z)(b_0', z') &= (b_0 b_0' + z b_0' + b_0 z', zz') \in \mathcal{A}_0; \\
(b_0, z)(b_1, 0) &= (b_0 b_1 + z b_1, 0) \in \mathcal{A}_1; \\
(b_1, 0)(b_0, z') &= (b_1 b_0 + b_1 z', 0) \in \mathcal{A}_1; \\
(b_1, 0)(b_1', 0) &= (b_1 b_1', 0) \in \mathcal{A}_0.
\end{aligned}
\tag{8.3}
$$

The results of (8.3) rely primarily on the fact that $\mathcal{B}$ is a superalgebra. We may conclude from (8.3) that the following inclusions hold:

$$
\mathcal{A}_0 \mathcal{A}_0, \ \mathcal{A}_1 \mathcal{A}_1 \subset \mathcal{A}_0; \qquad \mathcal{A}_0 \mathcal{A}_1, \ \mathcal{A}_1 \mathcal{A}_0 \subset \mathcal{A}_1.
\tag{8.4}
$$

From (8.4) we may conclude that $\mathcal{A}$ is a superalgebra with even subspace $\mathcal{A}_0$ and odd subspace $\mathcal{A}_1$. Thus we have that $\mathcal{A}$ is a unital associative $\dagger$-superalgebra and, as such, we may form the Lie $\dagger$-superalgebra $\mathcal{A}_{SLie}$. We now show that the map $\phi: \mathcal{L} \to \mathcal{A}_{SLie}$ defined by $L \mapsto (\phi_1(L), 0)$ is a Lie $\dagger$-superalgebra morphism. We have that for arbitrary homogeneous $L_1, L_2 \in \mathcal{L}$

$$
\begin{aligned}
&\{\phi(L_1), \phi(L_1)\}_{\mathcal{A}} \\
={} &\{(\phi_1(L_1), 0), (\phi_1(L_2), 0)\}_{\mathcal{A}} \\
={} &(\phi_1(L_1), 0)(\phi_1(L_2), 0) - (-1)^{\sigma((\phi_1(L_1), 0))\sigma((\phi_1(L_2), 0))}(\phi_1(L_2), 0)(\phi_1(L_1), 0).
\end{aligned}
\tag{8.5}
$$

From the above discussion concerning the grading of $\mathcal{A}$ we have that

$$
\sigma((\phi_1(L_i), 0)) = \sigma(\phi_1(L_i))
$$

for $i = 1, 2$. Implementing the definition of the multiplication in $\mathcal{A}$ given by (8.1) we re-write (8.5) as

$$
\begin{aligned}
&(\phi_1(L_1)\phi_1(L_2), 0) - (-1)^{\sigma(\phi_1(L_1))\sigma(\phi_1(L_2))}(\phi_1(L_2)\phi_1(L_1), 0) \\
={} &(\phi_1(L_1)\phi_1(L_2) - (-1)^{\sigma(\phi_1(L_1))\sigma(\phi_1(L_2))}\phi_1(L_2)\phi_1(L_1), 0) \\
={} &(\phi_1(\{L_1, L_2\}_{\mathcal{L}}), 0) \\
={} &\phi(\{L_1, L_2\}_{\mathcal{L}}).
\end{aligned}
$$

Thus ϕ is a Lie superalgebra morphism. For arbitrary $L \in \mathcal{L}$ we have from (8.2) that

$$
\phi(L^\dagger) = (\phi_1(L^\dagger), 0) = (\phi_1(L)^\dagger, 0) = (\phi_1(L), 0)^\dagger.
$$

So now we have that ϕ is a Lie $\dagger$-superalgebra morphism.

The universal property of $\mathcal{U}$ guarantees the existence of a $\dagger$-homomorphism $\tilde{\phi}: \mathcal{U} \to \mathcal{A}$ extending ϕ. For arbitrary $U \in \mathcal{U}$ we may write

$$\tilde{\phi}(U) = (\tilde{\phi}_1(U), \eta_1(U)).$$

Note that for arbitrary $L \in \mathcal{L}$ we have that $\tilde{\phi}(L) = \phi(L) = (\phi_1(L), 0)$. Thus, for all $L \in \mathcal{L}$ we have $\eta_1(L) = 0$. As η_1 is a unital †-superalgebra morphism $\mathcal{U} \to \mathbf{C}$ we have by the universality of $\mathcal{U}$ that η_1 is the unique map on $\mathcal{U}$ that sends each element of $\mathcal{L}$ to zero. Thus we have that η_1 is equal to η, the co-unit of $\mathcal{U}$. Therefore, given arbitrary $K \in \mathcal{K}$ we have $\tilde{\phi}(K) = (\tilde{\phi}_1(K), 0)$.

The restriction of $\tilde{\phi}_1$ to $\mathcal{K}$ is a †-superalgebra morphism from $\mathcal{K}$ to $\mathcal{B}$. This can be seen from the fact that, for arbitrary $K_1, K_2 \in \mathcal{K}$ we have

$$(\tilde{\phi}_1(K_1)\tilde{\phi}_1(K_2), 0) = (\tilde{\phi}_1(K_1), 0)(\tilde{\phi}_1(K_2), 0)$$
$$= \tilde{\phi}(K_1)\tilde{\phi}(K_2) = \tilde{\phi}(K_1 K_2) = (\tilde{\phi}_1(K_1 K_2), 0)$$

so that $\tilde{\phi}_1(K_1)\tilde{\phi}_1(K_2) = \tilde{\phi}_1(K_1 K_2)$. Furthermore, the restriction of $\tilde{\phi}_1$ to $\mathcal{K}$ extends ϕ_1 as we have for arbitrary $L \in \mathcal{L}$ that

$$(\tilde{\phi}_1(L), 0) = \tilde{\phi}(L, 0) = \phi(L, 0) = (\phi_1(L), 0)$$

so that $\tilde{\phi}_1(L) = \phi_1(L)$.

We have, then, established that any Lie †-superalgebra morphism $\phi_1 \colon \mathcal{L} \to \mathcal{B}$ may be extended to $\mathcal{K}$. It remains to establish that this extension is unique.

Suppose $\tilde{\psi}_1$ is a second †-homomorphism extending ϕ_1. By corollary 3.7, every element of $\mathcal{U}$ can be written uniquely as $K + z1$ with $K \in \mathcal{K}$ and $z \in \mathbf{C}$. This enables us to define a map $\tilde{\psi} \colon \mathcal{U} \to \mathcal{A}$ by $\tilde{\psi}(K + z1) := (\tilde{\psi}_1(K), \eta(z1))$. For arbitrary $L \in \mathcal{L}$ we have $\tilde{\psi}(L) = (\tilde{\psi}_1(L), \eta(0)) = (\phi_1(L), 0) = \phi(L)$ so $\tilde{\psi}$ is certainly an extension of ϕ. Furthermore, for arbitrary $U \in \mathcal{U}$ with $U = K + z1$, $K \in \mathcal{K}$ and $z \in \mathbf{C}$ we have

$$\tilde{\psi}(U^\dagger) = \tilde{\psi}((K + z1)^\dagger) = (\tilde{\psi}_1(K^\dagger), \eta(\bar{z}1))$$
$$= (\tilde{\psi}_1(K)^\dagger, \eta(\bar{z}1)) = (\tilde{\psi}_1(K), \eta(z1))^\dagger = \tilde{\psi}(U)^\dagger.$$

Finally we have for arbitrary U, V in $\mathcal{U}$ with $U = K + z1$ and $V = K' + z'1$ where $K, K' \in \mathcal{K}$ and $z, z' \in \mathbf{C}$ that

$$\tilde{\psi}(U)\tilde{\psi}(V) = (\tilde{\psi}_1(K), \eta(z1))(\tilde{\psi}_1(K'), \eta(z'1))$$
$$= (\tilde{\psi}_1(K)\tilde{\psi}_1(K') + z\tilde{\psi}_1(K') + z'\tilde{\psi}_1(K), \eta(zz'1))$$
$$= (\tilde{\psi}_1(KK' + zK' + z'K), \eta(zz'1))$$
$$= \tilde{\psi}(UV).$$

Thus $\tilde{\psi}$ is a †-superalgebra morphism extending ϕ and, by universality, this extension is unique. Thus $\tilde{\psi} = \tilde{\phi}$ from which it follows that $\tilde{\psi}_1 = \tilde{\phi}_1$ giving the required uniqueness of $\tilde{\phi}_1$.∎

8.3 The Co-Product and the Difference Map κ

We claim that the map $g: \mathcal{L} \to (\mathcal{U} \otimes \mathcal{U})_{SLie}$ defined by $g: L \mapsto L \otimes 1 + 1 \otimes L$ is a Lie †-superalgebra morphism. Firstly, for arbitrary $L \in \mathcal{L}$ we have that $g(L^\dagger) = L^\dagger \otimes 1 + 1 \otimes L^\dagger = (L \otimes 1 + 1 \otimes L)^\dagger = g(L)^\dagger$. Note that the fact $\sigma(1) = 0$ means that no grading factors appear in this calculation. We have, therefore, that g is a †-morphism as defined in Section 3.2. It is clear that $\sigma(L \otimes 1 + 1 \otimes L) = \sigma(L) + \sigma(1) = \sigma(L)$ so we have that g is parity-preserving. Finally, for arbitrary homogeneous $L_1, L_2 \in \mathcal{L}$ we have the standard result

$$
\begin{aligned}
&\{g(L_1), g(L_2)\}_{(\mathcal{U} \otimes \mathcal{U})_{SLie}} \\
=&\{L_1 \otimes 1 + 1 \otimes L_1, L_2 \otimes 1 + 1 \otimes L_2\}_{(\mathcal{U} \otimes \mathcal{U})_{SLie}} \\
=&L_1 L_2 \otimes 1 + (-1)^{\sigma(L_1)\sigma(L_2)} L_2 \otimes L_1 + L_1 \otimes L_2 + 1 \otimes L_1 L_2 \\
&-(-1)^{\sigma(L_1)\sigma(L_2)}(L_2 L_1 \otimes 1 + L_2 \otimes L_1 + (-1)^{\sigma(L_1)\sigma(L_2)} L_1 \otimes L_2 + 1 \otimes L_1 L_2) \\
=&L_1 L_2 \otimes 1 + 1 \otimes L_1 L_2 - (-1)^{\sigma(L_1)\sigma(L_2)}(L_2 L_1 \otimes 1 + 1 \otimes L_2 L_1) \\
=&(L_1 L_2 - (-1)^{\sigma(L_1)\sigma(L_2)} L_2 L_1) \otimes 1 + 1 \otimes (L_1 L_2 - (-1)^{\sigma(L_1)\sigma(L_2)} L_2 L_1) \\
=&g(\{L_1, L_2\}_{\mathcal{L}}).
\end{aligned}
$$

The equality between the penultimate and last lines holds because it is an intrinsic property of $\mathcal{U}$ that, for arbitrary homogeneous $L_1, L_2 \in \mathcal{L}$, the element $L_1 L_2 - (-1)^{\sigma(L_1)\sigma(L_2)} L_2 L_1$ of $\mathcal{U}$ is identified (via $\imath$) with $\{L_1, L_2\}_{\mathcal{L}}$.

We now have that g is a Lie †-superalgebra morphism from $\mathcal{L}$ to $(\mathcal{U} \otimes \mathcal{U})_{SLie}$ and, as such, the universal property guarantees the existence of an extension of g to the whole of $\mathcal{U}$. We denote this map from $\mathcal{U}$ to $\mathcal{U} \otimes \mathcal{U}$ by γ. This map is known as the *co-product*.

The map γ will now be used to define another map from $\mathcal{U}$ to $\mathcal{U} \otimes \mathcal{U}$. For an arbitrary element U of $\mathcal{U}$ we define the difference map $\kappa: \mathcal{U} \to \mathcal{U} \otimes \mathcal{U}$ by

$$
\kappa(U) = \gamma(U) - U \otimes 1.
$$

Note that κ is not a superalgebra morphism. This can be seen by taking some $L, M \in \mathcal{L}$ and observing that

$$
\kappa(L)\kappa(M) = (1 \otimes L)(1 \otimes M) = 1 \otimes LM
$$

whereas

$$
\begin{aligned}
\kappa(LM) =&(L \otimes 1 + 1 \otimes L)(M \otimes 1 + 1 \otimes M) - LM \otimes 1 \\
=&L \otimes M + (-1)^{\sigma(L)\sigma(M)} M \otimes L + 1 \otimes LM \neq \kappa(L)\kappa(M).
\end{aligned}
$$

It is, however, clear that κ is parity-preserving. Note also that $\kappa(1) = 0$.

Recall from Section 3.4 that any element U of $\mathcal{U}$ may be assigned a degree, $\deg U$, which is associated with the decomposition of U into basis elements as described by the Poincaré-Birkoff-Witt theorem stated in theorem 3.6. We use this notion of degree and the Poincaré-Birkoff-Witt theorem itself in the proposition that follows. Recall that X denotes an index set for a basis $(L_i)_{i \in X}$ of $\mathcal{L}$ consisting of arbitrary homogeneous elements, that $\tilde{X}$ denotes the set of admissible finite sequences $(i_1, \ldots, i_s)$ in X and that, given $\alpha = (i_1, \ldots, i_s) \in \tilde{X}$, we denote by L_α the element $L_{i_1} \cdots L_{i_s}$ of $\mathcal{U}$.

Proposition 8.4 The morphism κ maps $\mathcal{U}$ into $\mathcal{U} \otimes \mathcal{K}$. Furthermore, κ possesses the property that, given some $\alpha \in \tilde{X}$, there exist $\beta_1, \ldots, \beta_m \in \tilde{X}$ with each $|\beta_j| < |\alpha|$ and elements $k_1, \ldots, k_m \in \mathcal{K}$ such that

$$\kappa(L_\alpha) = \sum_{j=1}^{m} L_{\beta_j} \otimes k_j.$$

The second property of κ described in proposition 8.4 will be referred to as the *degree reducing property*.

Proof. Suppose $\alpha = (i_1, \ldots, i_r)$. Then we have

$$\kappa(L_\alpha) = (L_{i_1} \otimes 1 + 1 \otimes L_{i_1}) \cdots (L_{i_r} \otimes 1 + 1 \otimes L_{i_r}) - L_{i_1} \ldots L_{i_r} \otimes 1. \quad (8.6)$$

The expansion of the right-hand side of (8.6) consists of $2^r - 1$ terms, each of which is a twofold product tensor. It is clear that the first entry of each of these terms must be of degree at most $r - 1$. This provides the required degree-reducing property. It is equally clear that the second entry of each term must be of degree at least 1 and so must be an element of $\mathcal{K}$. Thus we have that $\kappa(L_\alpha) \in \mathcal{U} \otimes \mathcal{K}$ and the proposition holds.∎

Denote the identity map from $\mathcal{U}$ to $\mathcal{U}$ by id. We may define maps $\kappa_n : \mathcal{U} \to \mathcal{U} \otimes \mathcal{K} \otimes \cdots \otimes \mathcal{K}$ (n copies of $\mathcal{K}$) for $n \geq 1$ recursively by

$$\kappa_1 := \kappa, \qquad \kappa_n := (\kappa_1 \otimes \underbrace{id \otimes \cdots \otimes id}_{n-1 \text{ times}}) \circ \kappa_{n-1}.$$

It is convenient to set $\kappa_0 = id$. Since each application of κ is degree-reducing we have that for each $U \in \mathcal{U}$ there exists an integer n with $n \geq 1$ such that $\kappa_n(U) = 0$.

Proposition 8.5 For all $n \geq 0$ we have

$$\kappa_n \otimes id \circ \kappa = \kappa_{n+1}.$$

Proof. We prove the proposition for $n \geq 1$ by induction. If $n = 1$ then we have $\kappa_1 \otimes id \circ \kappa = \kappa_2$ by the definition of κ_2. Thus we have the base for the induction. Suppose the proposition holds for all $n < k$ where k is some integer. Then for $n = k$ we have

$$\kappa_k \otimes id \circ \kappa$$
$$= (\kappa_1 \otimes \underbrace{id \otimes \cdots \otimes id}_{k-1 \text{ times}} \circ \kappa_{k-1}) \otimes id \circ \kappa$$
$$= (\kappa_1 \otimes \underbrace{id \otimes \cdots \otimes id}_{k \text{ times}}) \circ (\kappa_{k-1} \otimes id) \circ \kappa$$

which, by the inductive hypothesis, is equal to

$$\kappa_1 \otimes \underbrace{id \otimes \cdots \otimes id}_{k \text{ times}} \circ \kappa_k$$

which, by definition, is equal to κ_{k+1}. Thus the inductive step is established and the proposition holds for all $n \geq 1$. For $n = 0$ it is clear that $id \otimes id \circ \kappa = \kappa = \kappa_1$. The proposition now holds.∎

Let ξ be the complex associative †-superalgebra morphism from $\mathcal{K}$ to $\mathcal{I}_{\text{SLie}}$ that extends the Lie superalgebra morphism $L \mapsto d\Xi_L$. The existence of ξ is guaranteed by proposition 8.3. For $n \geq 0$ consider the morphism $\eta \otimes \xi \otimes \cdots \otimes \xi$ (n copies of ξ). The domain of this map is $\mathcal{U} \otimes \mathcal{K} \otimes \cdots \otimes \mathcal{K}$ (n copies of $\mathcal{K}$) and its co-domain is $\mathbf{C} \otimes \mathcal{I} \otimes \cdots \otimes \mathcal{I}$ (n copies of $\mathcal{I}$). We may identify the range of $\eta \otimes \xi \otimes \cdots \otimes \xi$ (n copies of ξ) with $\mathcal{I} \otimes \cdots \otimes \mathcal{I}$ (n copies of $\mathcal{I}$). This family of morphisms plays an important part in the following section.

8.4 The Chaos Map

In this section we state and prove a theorem that gives an explicit formula for the map J given in corollary 8.2.

Theorem 8.6 If $\chi \colon \mathcal{U} \to S(\mathcal{I})$ is defined component-wise for $n \geq 0$ by

$$\chi(U)_n = \eta \otimes \underbrace{\xi \otimes \cdots \otimes \xi}_{n \text{ times}} \circ \kappa_n(U), \tag{8.7}$$

where U is an arbitrary element of $\mathcal{U}$, then $J(U) = I(\chi(U))$.

The map χ is known as the *chaos map*.

Proof. Let $\tilde{J}$ be defined to be the map $\mathcal{U} \to \mathcal{P}$ that sends an arbitrary element U of $\mathcal{U}$ to $I(\chi(U))$. Our aim is to show that $J = \tilde{J}$. The uniqueness guaranteed by the universal property of $\mathcal{U}$ means that this will be the case if it can be shown that $\tilde{J}$ satisfies

(i) $\tilde{J}(L) = \Xi_L$ for each $L \in \mathcal{L}$;

(ii) $\tilde{J}(U) \odot \tilde{J}(V) = \tilde{J}(UV)$ for all $U, V \in \mathcal{U}$.

It is a straightforward matter to establish (i). Consider an arbitrary element L of $\mathcal{L}$. From the definition of χ we have that $\chi(L)_0 = \eta(L) = 0$ and $\chi(L)_1 = \eta \otimes \xi \circ \kappa_1(L) = \eta \otimes \xi(1 \otimes L) = d\Xi_L$. For $n > 1$ we have that $\kappa_n(L) = 0$ so that for each $n > 1$ we have $\chi(L)_n = 0$. Therefore $\tilde{J}(L) = I(\chi(L)) = I((0, d\Xi_L, 0, \ldots)) = \Xi_L$ as required.

To establish (ii) we establish an analogue of the classical relation

$$df(\Lambda) = f(\Lambda + d\Lambda) - f(\Lambda) \tag{8.8}$$

where f is a polynomial function. A rigorous integral form of (8.8) in our notation would be

$$\tilde{J}(U) = \eta(U)Id + \int \tilde{J} \otimes \xi \circ \kappa(U). \tag{8.9}$$

The proof of (ii) begins by establishing that the relation in (8.9) does in fact hold. First we note that $\int \tilde{J} \otimes \xi \circ \kappa(U)$ may be re-written as $I(\chi \otimes \xi \circ \kappa(U))$. The injectivity of the map I established in proposition 7.8 means that the problem is reduced to establishing that

$$\chi(U) = \eta(U) + \chi \otimes \xi \circ \kappa(U). \tag{8.10}$$

The 0th order component of $\chi(U)$ is $\eta \circ \kappa_0(U) = \eta(U)$ which is also the 0th order component of the right-hand side of (8.10). For $n \geq 1$, the nth order component of $\chi(U)$ is $\eta \otimes \xi \otimes \cdots \otimes \xi \circ \kappa_n(U)$ (n copies of ξ). On the right-hand side of (8.10) the nth order component is

$$(\underbrace{\eta \otimes \xi \otimes \cdots \otimes \xi}_{n-1 \text{ times}} \circ \kappa_{n-1}) \otimes \xi \circ \kappa(U). \tag{8.11}$$

The rightmost ξ in (8.11) operates on the second entry of the twofold tensor $\kappa(U)$ which we recall to be an element of $\mathcal{U} \otimes \mathcal{K}$. On the other hand, the map $\eta \otimes \xi \otimes \cdots \otimes \xi \circ \kappa_{n-1}$ of (8.11) operates on the first entry of $\kappa(U)$. Thus we may delay the operation of ξ by one composition and include it in the leftmost map of (8.11). This manipulation re-writes (8.11) as

$$(\underbrace{\eta \otimes \xi \otimes \cdots \otimes \xi}_{n \text{ times}}) \circ \kappa_{n-1} \otimes id \circ \kappa(U).$$

By proposition 8.5 this is equal to

$$(\underbrace{\eta \otimes \xi \otimes \cdots \otimes \xi}_{n \text{ times}}) \circ \kappa_n(U)$$

which is the nth order component of $\chi(U)$ given in (8.7). Having proved relation (8.10) we may conclude that (8.9) indeed holds by proposition 7.8. We now proceed with the main part of the proof of (ii).

We proceed by induction on the degree of UV which is equal to $\deg U + \deg V$. If at least one of U, V is of degree zero then the result is immediate.

This is because, for an arbitrary $z \in \mathbf{C}$, we have $J(z1) = zId = I(z) = I(\eta(z1)) = I(\chi(z1)) = \tilde{J}(z1)$. If $\deg U = \deg V = 1$ then, by corollary 3.7, U may be written as $z1 + L$ with $z \in \mathbf{C}$ and $L \in \mathcal{L}$. Similarly, V may be written as $w1 + M$ with $w \in \mathbf{C}$ and $M \in \mathcal{L}$. Linearity allows us to assume that L and M are of definite parity. By (8.9) and (4.17) we have

$$\tilde{J}(U) = zId + \Xi_L, \qquad \tilde{J}(V) = wId + \Xi_M$$

which means that $\tilde{J}(U) \odot \tilde{J}(V)$ is equal to

$$zwId + z\Xi_M + w\Xi_L + \Xi_L \odot \Xi_M. \tag{8.12}$$

On the other hand, the $\mathcal{U}$ product UV is equal to $(z1 + L)(w1 + M)$ which is the same as

$$zw1 + wL + zM + LM. \tag{8.13}$$

It is easily shown from the definition given by (8.7) that

$$\chi(wL) = (0, wd\Xi_L, 0, \ldots)$$

and

$$\chi(zM) = (0, zd\Xi_M, 0, \ldots).$$

For the term $zw1$ of (8.13) we have

$$\chi(zw1)_0 = \eta(zw1) = zw$$

and that $\chi(zw1)_n = 0$ for all $n > 0$. For the $\mathcal{U}$ product LM we have

$$\begin{aligned}
\chi(LM)_0 &= \eta_0\kappa_0(LM) = \eta(LM) = 0 \\
\chi(LM)_1 &= \eta \otimes \xi \circ \kappa_1(LM) \\
&= \eta \otimes \xi((L \otimes 1 + 1 \otimes L)(M \otimes 1 + 1 \otimes M) - LM \otimes 1) \\
&= \eta \otimes \xi(L \otimes M + (-1)^{\sigma(L)\sigma(M)} M \otimes L + 1 \otimes LM) \\
&= \xi(LM) \\
&= d\Xi_{LM} \\
\chi(LM)_2 &= \eta \otimes \xi \otimes \xi \circ \kappa_2(LM) \\
&= \eta \otimes \xi \otimes \xi \circ \kappa_1 \otimes id \circ \kappa_1(LM) \\
&= \eta \otimes \xi \otimes \xi \circ \kappa_1 \otimes id(L \otimes M + (-1)^{\sigma(L)\sigma(M)} M \otimes L + 1 \otimes LM) \\
&= \eta \otimes \xi \otimes \xi \circ (1 \otimes L \otimes M + (-1)^{\sigma(L)\sigma(M)} 1 \otimes M \otimes L) \\
&= \xi(L) \otimes \xi(M) + (-1)^{\sigma(L)\sigma(M)} \xi(M) \otimes \xi(L) \\
&= d\Xi_L \otimes d\Xi_M + (-1)^{\sigma(L)\sigma(M)} d\Xi_M \otimes d\Xi_L.
\end{aligned}$$

Combining these calculations we have

$$\begin{aligned}
\chi(UV) = (zw, wd\Xi_L + zd\Xi_M + d\Xi_{LM}, d\Xi_L \otimes d\Xi_M + \\
(-1)^{\sigma(L)\sigma(M)} d\Xi_M \otimes d\Xi_L, 0, 0, \ldots)
\end{aligned}$$

so that

$$\tilde{J}(UV) = I(zw + wd\Xi_L + zd\Xi_M + d\Xi_{LM} + \\ d\Xi_L \otimes d\Xi_M + (-1)^{\sigma(L)\sigma(M)} d\Xi_M \otimes d\Xi_L). \tag{8.14}$$

The definition of the $\star$ and $\odot$ products allows us to re-write (8.14) as

$$\tilde{J}(UV) = I(zw + zd\Xi_M + wd\Xi_L + d(\Xi_L \odot \Xi_M)). \tag{8.15}$$

By the definition of I we have that (8.15) is equal to

$$zw\mathrm{Id} + z\Xi_L + w\Xi_M + \Xi_L \odot \Xi_M$$

which is equal to

$$\tilde{J}(U) \odot \tilde{J}(V)$$

as required.

We now have that (ii) holds for all $U, V \in \mathcal{U}$ so that $\deg U + \deg V \leq 2$. Assume, by way of induction, that the result holds for all $U, V \in \mathcal{U}$ such that $\deg U + \deg V < k$ for some positive integer k. Now take $U, V \in \mathcal{U}$ such that $\deg U + \deg V = k$. We must show that $\tilde{J}(U) \odot \tilde{J}(V) = \tilde{J}(UV)$. Consider the processes $\tilde{J}(U), \tilde{J}(V)$ at time zero. From (8.9) we have that $\tilde{J}(U)_0 = \eta(U)\mathrm{Id}$ and $\tilde{J}(V)_0 = \eta(V)\mathrm{Id}$. Also from (8.9) we have that the product $\tilde{J}(U) \odot \tilde{J}(V)$ is equal to

$$(\eta(U)\mathrm{Id} \odot \eta(V)\mathrm{Id})$$
$$+ (\eta(U)\mathrm{Id}) \odot \left(\int \tilde{J} \otimes \xi \circ \kappa(V) \right)$$
$$+ \left(\int \tilde{J} \otimes \xi \circ \kappa(U) \right) \odot (\eta(V)\mathrm{Id}) \tag{8.16}$$
$$+ \left(\int \tilde{J} \otimes \xi \circ \kappa(U) \right) \odot \left(\int \tilde{J} \otimes \xi \kappa(V) \right).$$

The definition of $\odot$ enables us to re-write (8.16) as

$$\eta(U)\eta(V)\mathrm{Id} + \eta(U) \int \tilde{J} \otimes \xi \circ \kappa(V) + \eta(V) \int \tilde{J} \otimes \xi \circ \kappa(U)$$
$$+ \int \left(\tilde{J} \otimes \xi \circ \kappa(U) \right) \star \left(\tilde{J} \otimes \xi \circ \kappa(V) \right). \tag{8.17}$$

Taking the expression (8.17) at time 0 gives us that $(\tilde{J}(U) \odot \tilde{J}(V))_0 = \eta(U)\eta(V)\mathrm{Id} = \eta(UV)\mathrm{Id}$. On the other hand, (8.9) gives us that $\tilde{J}(UV)_0 = \eta(UV)\mathrm{Id}$. Thus we may conclude that $(J(U) \odot J(V))_0 = J(UV)_0$.

The theorem will now be proved if we can show that the following differential relation holds:

$$d(\tilde{J}(U) \odot \tilde{J}(V)) = d\tilde{J}(UV). \tag{8.18}$$

Applying relation (7.36) to the left-hand side of (8.18) gives

$$d(\tilde{J}(U) \odot \tilde{J}(V)) = \tilde{J}(U)\Diamond d\tilde{J}(V) + d\tilde{J}(U)\Diamond\tilde{J}(V) + d\tilde{J}(U)d\tilde{J}(V). \quad (8.19)$$

If W is an arbitrary element of $\mathcal{U}$ then we may write (8.9) in a differential form as

$$d\tilde{J}(W) = \tilde{J} \otimes \xi \circ \kappa(W). \quad (8.20)$$

Putting (8.20) into (8.19) yields

$$\begin{aligned} d(\tilde{J}(U) \odot \tilde{J}(V)) = &\tilde{J}(U)\Diamond(\tilde{J} \otimes \xi \circ \kappa(V)) \\ &+ (\tilde{J} \otimes \xi \circ \kappa(U))\Diamond\tilde{J}(V) \\ &+ \left(\tilde{J} \otimes \xi \circ \kappa(U)\right)\left(\tilde{J} \otimes \xi \circ \kappa(V)\right). \end{aligned} \quad (8.21)$$

The degree-reducing property of κ allows us to invoke the inductive hypothesis and re-write the right-hand side of (8.21) as

$$\tilde{J} \otimes \xi \left((U \otimes 1)\kappa(V) + \kappa(U)(V \otimes 1) + \kappa(U)\kappa(V)\right). \quad (8.22)$$

Applying the definition of κ we re-write (8.22) as

$$\begin{aligned} \tilde{J} \otimes \xi \big((U \otimes 1)(\gamma(V) - V \otimes 1) \\ + (\gamma(U) - U \otimes 1)(V \otimes 1) \\ + (\gamma(U) - U \otimes 1)(\gamma(V) - V \otimes 1)\big) \end{aligned}$$

which, by simple cancellation of terms, is equal to

$$\tilde{J} \otimes \xi \left(\gamma(U)\gamma(V) - UV \otimes 1\right)$$

which, by the morphism property of γ, is the same as

$$\tilde{J} \otimes \xi \left(\gamma(UV) - UV \otimes 1\right)$$

which, by the definition of κ, is equal to

$$\tilde{J} \otimes \xi \circ \kappa(UV)$$

which, by (8.20), is equal to

$$d\tilde{J}(UV)$$

as required.∎

The ungraded analogue of this proof can be compared with the combinatorial proof of the corresponding theorem in [HPu].

9 Extensions

9.1 Introduction

In this chapter we explore some extensions of the theory that has now been covered. Sections 9.2 to 9.8 are concerned with the development of a theory of $\mathbf{Z}_n$-graded quantum stochastic calculus. While this theory does not yield useful commutation relations, a preliminary example of a $\mathbf{Z}_n \times \mathbf{Z}_n$-grading of quantum stochastic calculus is given in Section 9.9 which provides commutation relations of the form $[\tilde{\Lambda}_A, \tilde{\Lambda}_{A'}]_\omega = \tilde{\Lambda}_{[A,A']_\omega}$. In Section 9.10 we present a $\mathbf{Z}_n$-graded quantum stochastic calculus with an infinite number of degrees of freedom.

9.2 $\mathbf{Z}_n$-Grading of $M_0(N)$

A $\mathbf{Z}_n$-grading of an algebra A is a decomposition of A into a family $(A_\gamma)_{\gamma \in \mathbf{Z}_n}$ of subspaces of A so that

$$A = \sum_{\gamma \in \mathbf{Z}_n} A_\gamma$$

and which, for arbitrary $\gamma, \gamma' \in \mathbf{Z}_n$, satisfies the inclusion

$$A_\gamma A_{\gamma'} \subset A_{\gamma+\gamma'}. \tag{9.1}$$

The addition in (9.1) is modulo n, the usual addition in $\mathbf{Z}_n$. In this chapter the mode of addition in force will always be clear contextually and so we choose to omit the clumsy 'mod n' throughout. Similarly, we will rely on context for the interpretation of $-\gamma$ in the $\mathbf{Z}_n$ sense. Details of algebras graded over arbitrary abelian groups may be found in [B,C,Sch2].

Recall that $M_0(N)$ is the associative algebra of all $(N+1) \times (N+1)$ matrices with entries in $\mathbf{C}$. The rows and columns of these matrices are indexed from 0 to N. The algebra has basis $\{E_\beta^\alpha : 0 \leq \alpha, \beta \leq N\}$ where E_β^α denotes the matrix that is zero everywhere except for a 1 in the αth column at the βth row. The multiplication in this algebra is defined for arbitrary $A, B \in M_0(N)$ by

$$A.B = A((\hat{\delta}))B$$

where standard matrix multiplication is in force on the right and $((\hat{\delta})) = ((\hat{\delta}^\alpha_\beta))_{\substack{0 \leq \alpha \leq N \\ 0 \leq \beta \leq N}}$ as described in Chapter 2. In particular, the product of a pair of basis elements E^α_β, E^γ_δ is $\hat{\delta}^\alpha_\delta E^\gamma_\beta$.

We now fix, once and for all, integers $r_1, \ldots, r_{n-1}$ with $0 \leq r_1 \leq \cdots \leq r_{n-1} \leq N$. It is convenient to define r_0 to be -1 and r_n to be N.

For $\gamma = 0, \ldots, n-1$ define $M_0(N)_\gamma$ by

$$M_0(N)_\gamma = \mathrm{span}\{E^\alpha_\beta : \exists l, m \text{ such that } r_{l-1} < \alpha \leq r_l,$$
$$r_{m-1} < \beta \leq r_m \text{ and } \gamma = l - m\}.$$

Proposition 9.1 For arbitrary γ, γ' we have

$$M_0(N)_\gamma . M_0(N)_{\gamma'} \subset M_0(N)_{\gamma+\gamma'}.$$

Proof. By linearity it suffices to show that the result holds for the basis elements E^α_β. Suppose that $E^\alpha_\beta \in M_0(N)_\gamma$ and $E^\epsilon_\mu \in M_0(N)_{\gamma'}$. Then there exist integers l, m, l', m' such that the following relations hold:

$$r_{l-1} < \alpha \leq r_l, \qquad r_{l'-1} < \epsilon \leq r_{l'}, \qquad r_{m-1} < \beta \leq r_m, \qquad r_{m'-1} < \mu \leq r_{m'};$$

$$l - m = \gamma, \qquad l' - m' = \gamma'.$$

If $\hat{\delta}^\alpha_\mu = 0$ then the result is immediate as each component space $M_0(N)_\gamma$ of $M_0(N)$ contains the element $((0))$. Otherwise we have $\hat{\delta}^\alpha_\mu = 1$ and so $\alpha = \mu$. It follows immediately that $E^\alpha_\beta . E^\epsilon_\mu = E^\epsilon_\beta$ and $l = m'$. From this we may deduce that the grade of $E^\alpha_\beta . E^\epsilon_\mu$ is $l' - m = l' - l + m' - m = \gamma + \gamma'$. Thus $E^\alpha_\beta . E^\epsilon_\mu \in M_0(N)_{\gamma+\gamma'}$ and the result follows.∎

As in the $\mathbf{Z}_2$-graded case we require a function σ that maps an element to its grade. Thus we define

$$\sigma : \cup_{\gamma=0}^{n-1} M_0(N)_\gamma \to \mathbf{Z}_n$$

by $\sigma(A) = \gamma$ where γ is the unique element of $\mathbf{Z}_n$ so that $A \in M_0(N)_\gamma$. If this prescription is to be followed slavishly, σ is not well-defined on $((0)) \in M_0(N)$. This is not a problem of practical concern and we choose $\sigma(((0)))$ to be any value in $\mathbf{Z}_n$ we wish.

If E^α_β is a basis element of $M_0(N)$ then it is certainly of definite grade. As a shorthand for $\sigma(E^\alpha_\beta)$ we write σ^α_β. Note that, in the $\mathbf{Z}_n$ sense, $\sigma^\alpha_\beta = -\sigma^\beta_\alpha$ and $\sigma^\alpha_\beta + \sigma^\gamma_\alpha = \sigma^\gamma_\beta$. These two relations allow us to write $\sigma^\alpha_\beta = \sigma^\alpha_0 - \sigma^\beta_0$.

We denote the graded structure obtained by grading $M_0(N)$ in this way by $M_0(N, \mathbf{r})$ where $\mathbf{r} = (r_1, \ldots, r_{n-1})$. We denote the subspace of $M_0(N, \mathbf{r})$ consisting of homogeneous elements of degree γ by $M_0(N, \mathbf{r})_\gamma$ as would be expected.

The grading procedure for $M_0(N)$ can be generalised by using an arbitrary partition of $\{0, \ldots, N\}$ rather than the particular partitioning employed here. Such a generalisation is not illuminating in the N-dimensional case but is necessary for the Z_n-grading of quantum stochastic calculus with infinitely many degrees of freedom described in Section 9.10.

9.3 Z_n Grading of Quantum Stochastic Calculus

Let us now describe the means by which quantum stochastic calculus may be given a Z_n-grading using an index space of the form $M_0(N, \mathbf{r})$.

Let ω be an arbitrary primitive n^{th} root of -1. Let H denote the $N \times N$ matrix with zero for all off-diagonal entries and for which the ith diagonal entry is $\omega^{\sigma_0^i}$ where σ_0^i is the grade of the element E_0^i in $\underline{M_0(N, \mathbf{r})}$ as described in Section 9.2. We remark that the complex conjugate $\overline{\omega^{\sigma_\beta^\alpha}}$ of $\omega^{\sigma_\beta^\alpha}$ is equal to $\omega^{\sigma_\alpha^\beta}$.

The Z_n-grading process $(G(t))_{t \geq 0}$ on $\mathcal{E}$ is defined by its action on an arbitrary exponential vector $e(f)$ where $f \in L^2(R_+; \mathbf{C}^N)$ as follows:

$$G(t)e(f) = e(\chi_{[0,t]} f H + \chi_{(t,\infty)} f). \tag{9.2}$$

To grade operators and processes we require the grading operator $G(\infty)$ defined by

$$G(\infty)e(f) = e(fH)$$

where f is an arbitrary element of $L^2(\mathbf{R}_+; \mathbf{C}^N)$.

For arbitrary α, β with $0 \leq \alpha, \beta \leq N$, the differentials $d\tilde{\Lambda}_\beta^\alpha$ are defined in terms of this grading process thus:

$$d\tilde{\Lambda}_\beta^\alpha = G^{\sigma_\alpha^\beta} d\Lambda_\beta^\alpha.$$

If an element A of $M_0(N, \mathbf{r})$ is decomposed as $\lambda_\alpha^\beta E_\beta^\alpha$ then the corresponding quantum stochastic differential $d\tilde{\Lambda}_A$ is taken to be the sum $\lambda_\alpha^\beta d\tilde{\Lambda}_\beta^\alpha$. To each $d\tilde{\Lambda}_A$ there corresponds a quantum stochastic process $\tilde{\Lambda}_A$. This process is defined in the obvious way as

$$\tilde{\Lambda}_A = \int_0^{\cdot} Id \otimes d\tilde{\Lambda}_A(s).$$

As might be expected, $\tilde{\Lambda}_\beta^\alpha$ is defined to be the process $\tilde{\Lambda}_{E_\beta^\alpha}$.

9.4 Conjugation

We begin with some preliminary comments relating to the means by which a grading operator should be implemented. In the $\mathbf{Z}_2$-graded case, the grading operator $G(\infty)$ is self-inverse. As a result, the parity of a process X may be taken to be 0 if the product $G(\infty)XG(\infty)$ is equal to X and 1 if the product $G(\infty)XG(\infty)$ is equal to $-X$. The means of implementing the grading process $G(\infty)$ of $\mathbf{Z}_n$-graded quantum stochastic calculus to provide a grading of all relevant quantum stochastic processes is less clear.

It is instructive in this matter to examine further the structure of $M_0(N, \mathbf{r})$. The obvious choice for a grading matrix K for this $\mathbf{Z}_n$-graded algebra is

$$K = \sum_{\gamma=0}^{N} \omega^{\sigma_0^{\gamma}} E_\gamma^\gamma.$$

The conjugate $K^{-1} E_\beta^\alpha K$ of an arbitrary basis element E_β^α of $M_0(N, \mathbf{r})$ by K is then equal to

$$\left(\sum_{\gamma=0}^{N} \omega^{\sigma_\gamma^0} E_\gamma^\gamma \right) E_\beta^\alpha \left(\sum_{\gamma=0}^{N} \omega^{\sigma_0^{\gamma}} E_\gamma^\gamma \right).$$

A brief calculation show this product to be equal to $\omega^{\sigma_0^{\alpha} + \sigma_\beta^0} E_\beta^\alpha$ which is in turn equal to $\omega^{\sigma_\beta^{\alpha}} E_\beta^\alpha$.

We may deduce from this result that for an arbitrary homogeneous element A of $M_0(N, \mathbf{r})$, the conjugate $K^{-1}AK$ of A yields $\omega^{\sigma(A)} A$ where $\sigma(A)$ is the degree of the matrix A as discussed in Section 9.2. Thus the subspace $M_0(N, \mathbf{r})_\gamma$ of $M_0(N, \mathbf{r})$ might equally well be defined to be the space $\{ A \in M_0(N, \mathbf{r}) : K^{-1}AK = \omega^\gamma A \}$. This result indicates that quantum stochastic processes might be graded by means of the product $G^{-1}(\infty)XG(\infty)$ where G now denotes the grading function defined by (9.2). This approach is vindicated in the remainder of this section.

It is easy to see that the inverse $G^{-1}(\infty)$ of $G(\infty)$ is equal to $G(\infty)^{n-1}$. It is also easy to see that $G(\infty)$ is unitary and so $G(\infty)^\dagger = G^{-1}(\infty)$. It immediately follows that $(G^{\sigma_\beta^{\alpha}}(\infty))^\dagger$ is equal to $G^{\sigma_\alpha^{\beta}}(\infty)$. The the fact that $G(\infty)$ leaves the exponential domain $\mathcal{E}$ invariant and the fact that $G(\infty)$ is unitary and hence defined on all of $\Gamma(L^2(\mathbf{R}_+; \mathbf{C}^N))$ means that, if X is a not-necessarily bounded operator defined on $\mathcal{E}$, the neat operator products $XG(\infty)$ and $G(\infty)X$ may be interpreted rigorously.

The following proposition shows that if A is an arbitrary homogeneous element of $M_0(N, \mathbf{r})$ then the grade of $\tilde{\Lambda}_A$ as determined using conjugation by $G(\infty)$ is equal to the grade of A. In general, the operators constituting a process of the form $\tilde{\Lambda}_A$ do not leave the exponential domain $\mathcal{E}$ invariant. It is therefore necessary to express the conjugation in terms of adjoints and the inner product.

Proposition 9.2 For arbitray $t, u \geq 0$ with $u \geq t$ (including the case $u = \infty$) and arbitrary homogeneous $A \in M_0(N, \mathbf{r})$ we have the equality

$$G(u)^{-1} \tilde{\Lambda}_A(t) G(u) = \omega^{\sigma(A)} \tilde{\Lambda}_A(t).$$

Proof. By linearity we may assume that A is a basis element E_β^α of $M_0(N, \mathbf{r})$ and so $\tilde{\Lambda}_A = \tilde{\Lambda}_\beta^\alpha$ for some α, β with $0 \leq \alpha, \beta \leq N$. Take arbitrary $e(f), e(g)$ where f, g are some elements of $L^2(R_+; \mathbf{C}^N)$. The inner product $\langle e(f), G(u)^{-1} \tilde{\Lambda}_\beta^\alpha(t) G(u) e(g) \rangle$ is equal to $\langle G(u) e(f), \tilde{\Lambda}_\beta^\alpha(t) G(u) e(g) \rangle$ and, by the first fundamental formula stated in theorem 2.7, this is equal to

$$\int_0^t \omega^{\sigma_\beta^0} f_\beta(t_1) \omega^{\sigma_0^\alpha} g^\alpha(t_1) \langle G(u) e(f), G^{\sigma_\alpha^\beta}(t_1) G(u) e(g) \rangle \, dt_1.$$

The unitarity of $G(u)$ and an obvious extension of lemma 4.3 allows us to re-write this expression as

$$\int_0^t f_\beta(t_1) g^\alpha(t_1) \omega^{\sigma_\beta^\alpha} \langle e(f), G^{\sigma_\alpha^\beta}(t_1) e(g) \rangle \, dt_1.$$

Another application of the first fundamental formula shows that this is equal to

$$\langle e(f), \omega^{\sigma_\beta^\alpha} \tilde{\Lambda}_\beta^\alpha(t) e(g) \rangle$$

and so the result follows.■

It is clear, then, that this method of conjugation is suitable for the grading of general quantum stochastic processes. The degree of a quantum stochastic process $(X(t) : t \geq 0)$ is taken to be γ if, for all $t \geq 0$, we have $G^{-1}(\infty) X(t) G(\infty) = \omega^\gamma X(t)$. If such a conjugation relation leading to a grade of X exists then X is said to be *homogeneous*. Any process X may be expressed as a sum of at most n homogeneous processes. The grade of a homogeneous process X will be written $\sigma(X)$ and, as might be expected, we write $\sigma(\tilde{\Lambda}_\beta^\alpha)$ as σ_β^α.

We now wish to determine the grade of an arbitrary iterated quantum stochastic integral in this theory. This is done by means of the conjugation result given in the next proposition.

Proposition 9.3 For arbitrary $m \geq 1$, arbitrary homogeneous elements $A_1, \ldots, A_m$ of $M_0(N, \mathbf{r})$ and arbitrary $u, t \geq 0$ with $u \geq t$ (including the case $u = \infty$) we have

$$G(u)^{-1} \int_{0 < t_1 < \cdots < t_m < t} d\tilde{\Lambda}_{A_1}(t_1) \ldots d\tilde{\Lambda}_{A_m}(t_m) G(u)$$

$$= \omega^{\sum_{i=1}^m \sigma(A_i)} \int_{0 < t_1 < \cdots < t_m < t} d\tilde{\Lambda}_{A_1}(t_1) \ldots d\tilde{\Lambda}_{A_m}(t_m).$$

Proof. It has already been established in proposition 9.2 that the result holds for the case $m = 1$ so we may proceed by induction on m. Suppose the proposition holds for all $m < k$ for some integer k. Now take $m = k$. By linearity we may assume that the A_i are basis elements of $M_0(N, \mathbf{r})$ so that for each i we have $d\tilde{\Lambda}_{A_i} = d\tilde{\Lambda}_{\beta_i}^{\alpha_i}$. The first fundamental formula shows that, if f, g are arbitrary elements of $\mathbf{L}^2(\mathbf{R}_+; \mathbf{C}^N)$ then we have

$$\langle e(f), G(u)^{-1} \int_{0 < t_1 < \cdots < t_k < t} d\tilde{\Lambda}_{\beta_1}^{\alpha_1}(t_1) \ldots d\tilde{\Lambda}_{\beta_k}^{\alpha_k}(t_k) G(u) e(g) \rangle$$

$$= \int_0^t \omega^{\sigma_{\beta_k}^{\alpha_k}} f_{\beta_k}(s) g^{\alpha_k}(s) \langle G(u) e(f),$$

$$\int_{0 < t_1 < \cdots < t_{k-1} < t_k} d\tilde{\Lambda}_{\beta_1}^{\alpha_1}(t_1) \ldots d\tilde{\Lambda}_{\beta_{k-1}}^{\alpha_{k-1}}(t_{k-1}) G^{\sigma_{\alpha_k}^{\beta_k}}(t_k) G(u) e(g) \rangle \, dt_k.$$

Invoking the inductive hypothesis and an the obvious extension of lemma 4.3 we can see that this expression is equal to

$$\int_0^t \omega^{\sigma_{\beta_k}^{\alpha_k}} f_{\beta_k}(s) g^{\alpha_k}(s) \langle e(f),$$

$$\omega^{\sum_{i=1}^{k-1} \sigma_{\beta_i}^{\alpha_i}} \int_{0 < t_1 < \cdots < t_{k-1} < t_k} d\tilde{\Lambda}_{\alpha_1}^{\beta_1}(t_1) \ldots d\tilde{\Lambda}_{\beta_{k-1}}^{\alpha_{k-1}}(t_{k-1}) G^{\sigma_{\alpha_k}^{\beta_k}}(t_k) e(g) \rangle \, dt_k.$$

An application of the first fundamental formula enables this to be re-written as

$$\omega^{\sum_{i=1}^{k} \sigma_{\beta_i}^{\alpha_i}} \langle e(f), \int_{0 < t_1 < \cdots < t_k < t} d\tilde{\Lambda}_{A_1}(t_1) \ldots d\tilde{\Lambda}_{A_k}(t_k) e(g) \rangle.$$

The result now follows by induction and the totality of the exponential vectors in $\mathcal{E}$. ∎

We complete this section by giving a conjugation result for quantum stochastic integrals of arbitrary homogeneous processes.

Proposition 9.4 Let E be an arbitrary homogeneous process integrable by the arbitrary homogeneous integrator $d\tilde{\Lambda}_A$. Then for arbitrary $t, u \geq 0$ with $u \geq t$ (including the case $u = \infty$) we have

$$G^{-1}(u) \int_0^t E(s) \otimes d\tilde{\Lambda}_A(s) G(u) = \omega^{\sigma(E) + \sigma(A)} \int_0^t E(s) \otimes d\tilde{\Lambda}_A(s).$$

Proof. Without loss of generality we may assume that A is of the form E_β^α. Proceeding in the usual way, using exponential vectors $e(f), e(g)$ with f, g being arbitrary elements of $\mathbf{L}^2(\mathbf{R}_+; \mathbf{C}^N)$, the first fundamental formula and the obvious extension of lemma 4.3 we have

$$\langle e(f), G^{-1}(u) \int_0^t E(s) \otimes d\tilde{\Lambda}_\beta^\alpha(s) G(u) e(g) \rangle$$

$$= \langle G(u)e(f), \int_0^t E(s) G^{\sigma_\alpha^\beta}(s) \otimes d\Lambda_\beta^\alpha(s) G(u) e(g) \rangle$$

$$= \int_0^t \omega^{\sigma_\beta^0} f_\beta(s) \omega^{\sigma_0^\alpha} g^\alpha(s) \langle G(u)e(f), E(s) G^{\sigma_\alpha^\beta}(s) G(u) e(g) \rangle \, ds$$

$$= \omega^{\sigma(E)+\sigma_\beta^\alpha} \int_0^t f_\beta(s) g^\alpha(s) \langle e(f), E(s) G^{\sigma_\alpha^\beta}(s) e(g) \rangle \, ds$$

$$= \omega^{\sigma(E)+\sigma_\beta^\alpha} \langle e(f), \int_0^t E(s) \otimes d\tilde{\Lambda}_\beta^\alpha(s) e(g) \rangle$$

and so the result follows.∎

9.5 The Taking of Adjoints

It was seen in Section 7.4 that, when taking adjoints of iterated quantum stochastic integrals in the $\mathbf{Z}_2$-graded case, the introduction of a power of -1 determined by the parity of the integrators in question was required. In this section we see that an exactly analogous situation occurs in the $\mathbf{Z}_n$-graded case. We begin by looking at the adjoint of processes of the form $\tilde{\Lambda}_A$ where A is an arbitrary element of $M_0(N, \mathbf{r})$.

Proposition 9.5 Let A be an arbitrary homogeneous element of $M_0(N, \mathbf{r})$. If $\bar{A}^T$ denotes the conjugate transpose (that is, adjoint) of A, then for each $t \geq 0$ we have

$$(\tilde{\Lambda}_A(t))^\dagger = \tilde{\Lambda}_{\bar{A}^T}(t).$$

Proof. By linearity it suffices show that result holds for the case where A is an arbitrary basis element E_β^α of $M_0(N, \mathbf{r})$. From our definitions and the standard theory of quantum stochastic calculus we have that, given an arbitrary time $t \geq 0$, the adjoint $(\tilde{\Lambda}_\beta^\alpha(t))^\dagger$ of the operator $\tilde{\Lambda}_\beta^\alpha(t)$ is equal to

$$\left(\int_0^t G^{\sigma_\alpha^\beta}(s) \otimes d\Lambda_\beta^\alpha(s) \right)^\dagger$$

$$= \int_0^t (G^{\sigma_\alpha^\beta}(s))^\dagger \otimes d\Lambda_\alpha^\beta(s)$$

$$= \int_0^t G^{\sigma_\beta^\alpha}(s) \otimes d\Lambda_\alpha^\beta(s)$$

$$= \tilde{\Lambda}_\alpha^\beta.$$

Having established that the proposition holds for the basis elements E_β^α of $M_0(N, \mathbf{r})$ we may conclude that it is true for all A in $M_0(N, \mathbf{r})$. ∎

Proposition 9.5 shows that, for an arbitrary homogeneous element A of $M_0(N, \mathbf{r})$, we have

$$\sigma(\tilde\Lambda_{\bar A^T}) = -\sigma(\tilde\Lambda_A)$$

as would be expected from the $M_0(N, \mathbf{r})$ relation $\sigma(\bar A^T) = -\sigma(A)$. This observation may be generalised to an arbitrary quantum stochastic process X defined on $\mathcal{E}$ as is seen from the next proposition.

Proposition 9.6 If X is an arbitrary homogeneous quantum stochastic process defined on $\mathcal{E}$ then $\sigma(X^\dagger) = -\sigma(X)$.

Proof. Taking an arbitrary $t \geq 0$ we have

$$G^{-1}(\infty)X^\dagger(t)G(\infty)$$
$$=(G^{-1}(\infty)X(t)G(\infty))^\dagger$$
$$=(\omega^{\sigma(X)}X(t))^\dagger$$
$$=\omega^{-\sigma(X)}X^\dagger(t)$$

and so the proposition holds. ∎

The next proposition provides a formula for the adjoint of an arbitrary iterated quantum stochastic integral in the $\mathbf{Z}_n$-graded case.

Proposition 9.7 For arbitrary $m \geq 1$, arbitrary $t \geq 0$ and arbitrary $A_1, \ldots, A_m$ in $M_0(N, \mathbf{r})$ we have that

$$\left(\int_{0 < t_1 < \cdots < t_m < t} d\tilde\Lambda_{A_1}(t_1) \ldots d\tilde\Lambda_{A_m}(t_m) \right)^\dagger$$

is equal to

$$\omega^{\sum_{i<j} \sigma(A_i)\sigma(A_j)} \int_{0 < t_1 < \cdots < t_m < t} d\tilde\Lambda_{\bar A_1^T}(t_1) \ldots d\tilde\Lambda_{\bar A_m^T}(t_m).$$

Proof. The case $m = 1$ has already been established in proposition 9.5. Thus we may proceed by induction on m. Without loss of generality we may assume that each A_i is a basis element and so each $d\tilde\Lambda_{A_i}$ may be written as $d\tilde\Lambda_{\beta_i}^{\alpha_i}$. Suppose that the proposition is true for all $m < k$ for some integer k. Now take $m = k$. Then, using the inductive hypothesis

$$\left(\int_{0 < t_1 < \cdots < t_k < t} d\tilde\Lambda_{\beta_1}^{\alpha_1}(t_1) \ldots d\tilde\Lambda_{\beta_k}^{\alpha_k}(t_k) \right)^\dagger$$

may be re-written as

$$_\omega\Sigma_{i<j\leq k-1}\,\sigma_{\beta_i}^{\alpha_i}\sigma_{\beta_j}^{\alpha_j}\int_0^t G^{\alpha_k}_{\beta_k}(t_k)\int_{0<t_1<\cdots<t_k} d\tilde{\Lambda}_{\alpha_1}^{\beta_1}(t_1)\ldots d\tilde{\Lambda}_{\alpha_{k-1}}^{\beta_{k-1}}(t_{k-1})\,d\Lambda_{\alpha_k}^{\beta_k}(t_k).$$

Proposition 9.3 above shows that this expression is equal to

$$_\omega\sigma_{\beta_k}^{\alpha_k}\sum_{i=1}^{k-1}\sigma_{\beta_i}^{\alpha_i}\,_\omega\Sigma_{i<j\leq k-1}\,\sigma_{\beta_i}^{\alpha_i}\sigma_{\beta_j}^{\alpha_j}\int_{0<t_1<\cdots<t_k<t} d\tilde{\Lambda}_{\alpha_1}^{\beta_1}(t_1)\ldots d\tilde{\Lambda}_{\alpha_k}^{\beta_k}(t_k)$$

which, by rearrangement of the powers of ω, may be re-written as

$$_\omega\Sigma_{i<j}\,\sigma_{\beta_i}^{\alpha_i}\sigma_{\beta_j}^{\alpha_j}\int_{0<t_1\cdots<t_k<t} d\tilde{\Lambda}_{\alpha_1}^{\beta_1}(t_1)\ldots d\tilde{\Lambda}_{\alpha_k}^{\beta_k}(t_k)$$

and so the required result follows by linearity.∎

We complete this section with a formula for the adjoint of a general $\mathbf{Z}_n$-graded quantum stochastic integral.

Proposition 9.8: For an arbitrary homogeneous integrable process E integrable by the arbitrary homogeneous integrator $d\tilde{\Lambda}_A$ and an arbitrary time $t \geq 0$, we have that

$$\left(\int_0^t E(s)\otimes d\tilde{\Lambda}_A(s)\right)^\dagger = \omega^{\sigma(E)\sigma(A)}\int_0^t E^\dagger(s)\otimes d\tilde{\Lambda}_{\bar{A}^T}(s).$$

Proof. Using the standard theory of quantum stochastic calculus and the fact that the grade of E is defined by means of conjugation by G we have

$$\left(\int_0^t E(s)\otimes d\tilde{\Lambda}_A(s)\right)^\dagger$$

$$=\left(\int_0^t E(s)G^{-\sigma(A)}(s)\otimes d\Lambda_A(s)\right)^\dagger$$

$$=\int_0^t G^{\sigma(A)}(s)E^\dagger(s)\otimes d\Lambda_{\bar{A}^T}(s)$$

$$=\omega^{\sigma(E)\sigma(A)}\int_0^t E^\dagger(s)G^{\sigma(A)}(s)\otimes d\Lambda_{\bar{A}^T}(s)$$

$$=\omega^{\sigma(E)\sigma(A)}\int_0^t E^\dagger(s)\otimes d\tilde{\Lambda}_{\bar{A}^T}$$

as required.∎

9.6 The Second Fundamental Formula in Z_n-Graded Quantum Stochastic Calculus

This section develops ideas concerning the second fundamental formula in the context of Z_n-graded quantum stochastic calculus. We begin by taking a rigorous approach which will provide results enabling us to give a more perspicuous formulation.

Suppose that $t \geq 0$ is an arbitrary time, that A, B are homogeneous elements of $M_0(N, \mathbf{r})$ and that E, F are homogeneous integrable processes. We would like to take the product of the quantum stochastic integrals

$$\int_0^t E(s) \otimes d\tilde{\Lambda}_A(s) \quad \text{and} \quad \int_0^t F(s) \otimes d\tilde{\Lambda}_B(s)$$

but are prevented from doing so by considerations of boundedness. Instead we take the adjoint of the first integral and consider the inner product

$$\langle \omega^{\sigma(A)\sigma(E)} \int_0^t E^\dagger(s) \otimes d\tilde{\Lambda}_{\bar{A}^T}(s)e(f), \int_0^t F(s) \otimes d\tilde{\Lambda}_B(s)e(g) \rangle \qquad (9.3)$$

where f and g are arbitrary elements of $L^2(\mathbf{R}_+; \mathbf{C}^N)$.

We may assume without loss of generality that A and B are basis elements of $M_0(N, \mathbf{r})$. We may therefore relabel A as E_β^α and B as E_δ^γ. With this simplification, the second fundamental formula stated in theorem 2.8 gives us that (9.3) is equal to

$$\int_0^t f_\delta(s)g^\gamma(s)\langle \omega^{\sigma_\beta^\alpha\sigma(E)} \int_0^s E^\dagger(r)\,d\tilde{\Lambda}_\alpha^\beta(r)e(f), F(s)G^{\sigma_\gamma^\delta}(s)e(g) \rangle\,ds$$

$$+ \int_0^t f_\beta(s)g^\alpha(s)\langle \omega^{\sigma_\beta^\alpha\sigma(E)} E^\dagger(s)G^{\sigma_\beta^\alpha}(s)e(f), \int_0^s F(r)\,d\tilde{\Lambda}_\delta^\gamma(r)e(g) \rangle\,ds$$

$$+ \hat{\delta}_\delta^\alpha f_\beta(s)g^\gamma(s) \int_0^t \langle \omega^{\sigma_\beta^\alpha\sigma(E)} E^\dagger(s)G^{\sigma_\beta^\alpha}(s)e(f), F(s)G^{\sigma_\gamma^\delta}(s)e(g) \rangle\,ds.$$

This might be interpreted formally as

$$\int_0^t E(s)\,d\tilde{\Lambda}_\beta^\alpha(s) \int_0^t F(s)\,d\tilde{\Lambda}_\delta^\gamma(s)$$

$$= \int_0^t \int_0^s E(r)\,d\tilde{\Lambda}_\beta^\alpha(r)F(s)\,d\tilde{\Lambda}_\delta^\gamma(s)$$

$$+ \omega^{\sigma_\beta^\alpha(\sigma(F)+\sigma_\delta^\gamma)} \int_0^t E(s) \int_0^s F(r)\,d\tilde{\Lambda}_\delta^\gamma(r)\,d\tilde{\Lambda}_\beta^\alpha(s) \qquad (9.4)$$

$$+ \hat{\delta}_\delta^\alpha \omega^{\sigma_\beta^\alpha\sigma(F)} \int_0^t E(s)F(s)\,d\tilde{\Lambda}_\beta^\gamma(s).$$

In the remainder of this section we will develop a formal differential version of (9.4) by providing a $\mathbf{Z}_n$-graded analogue of the Chevalley tensor product and defining an action of $\mathcal{P}$ on $\mathcal{P} \otimes \mathcal{I}$.

It has already been seen in Chapter 3 that a salient feature of the Chevalley tensor product is that the multiplication of such tensors introduces a grading factor. The important role of this multiplication in the context of $\mathbf{Z}_2$-graded quantum stochastic calculus was seen in Section 7.3. It will be similarly useful to define a multiplication of tensor products of associative $\mathbf{Z}_n$-graded algebras by linear extension of the following rule for product tensors with homogeneous entries as follows:

$$(a_1 \otimes a_2)(b_1 \otimes b_2) = \omega^{\sigma(a_2)\sigma(b_1)} a_1 b_1 \otimes a_2 b_2.$$

It must be shown that this multiplication is associative. Let $a_1, a_2, b_1, b_2, c_1, c_2$ be of definite parity Then under this new multiplication we have

$$
\begin{aligned}
((a_1 \otimes a_2)(b_1 \otimes b_2))(c_1 \otimes c_2) &= \omega^{\sigma(a_2)\sigma(b_1)}(a_1 b_1 \otimes a_2 b_2)(c_1 \otimes c_2) \\
&= \omega^{\sigma(a_2)\sigma(b_1)+\sigma(c_1)(\sigma(a_2)+\sigma(b_2))} a_1 b_1 c_1 \otimes a_2 b_2 c_2 \\
&= \omega^{\sigma(c_1)\sigma(b_2)+\sigma(a_2)(\sigma(b_1)+\sigma(c_1))} a_1 b_1 c_1 \otimes a_2 b_2 c_2 \\
&= \omega^{\sigma(c_1)\sigma(b_2)}(a_1 \otimes a_2)(b_1 c_1 \otimes b_2 c_2) \\
&= (a_1 \otimes a_2)((b_1 \otimes b_2)(c_1 \otimes c_2)).
\end{aligned}
$$

Associativity follows by linearity.

As is conventional in this work, let $\mathcal{I}$ denote an Ito $\mathbf{Z}_n$-graded algebra of $\mathbf{Z}_n$-graded quantum stochastic calculus. This might consist of all the $d\tilde{\Lambda}_A$ or a $\mathbf{Z}_n$-graded subalgebra thereof. Let $\mathcal{P}$ denote a space of quantum stochastic processes integrable by the elements of $\mathcal{I}$. We might, for example, take $\mathcal{P}$ to be the space of all iterated $\mathbf{Z}_n$-graded quantum stochastic integrals.

We wish to define an action $\diamond$ of $\mathcal{P}$ on $\mathcal{P} \otimes \mathcal{I}$. The left and right actions of $\diamond$ are defined by linear extension of the following rules for arbitrary homogeneous elements E, F of $\mathcal{P}$ and an arbitrary homogeneous element dM of $\mathcal{I}$:

$$E \diamond (F \otimes dM) = EF \otimes dM$$

$$(F \otimes dM) \diamond E = \omega^{\sigma(dM)\sigma(E)} FE \otimes dM.$$

These expressions must be interpreted formally as they contain products of unbounded operators.

Suppose we have two quantum stochastic integrals

$$A = \int A_1 \otimes dA_2, \qquad B = \int B_1 \otimes dB_2$$

where A_1, A_2, B_1, B_2 are all of definite grade. Then, using the notation just described, we may express the second fundamental formula as

$$dAB = A \diamond (B_1 \otimes dB_2) + (A_1 \otimes dA_2) \diamond B + (A_1 \otimes dA_2)(B_1 \otimes dB_2).$$

The formal products in use here are interpreted rigorously by means of adjoints and the inner product.

9.7 The Higher Order Ito Product Formula in Z_n-Graded Quantum Stochastic Calculus

Consider the space $T(\mathcal{I})$ of all tensors of finite rank formed from $\mathcal{I}$. In this section $\mathcal{I}$ is taken to be either the Ito Z_n-graded algebra consisting of all the $d\tilde{\Lambda}_A$ or a Z_n-graded subalgebra thereof. We define an involution on $T(\mathcal{I})$ by linear extension of the following rule for product tensors with entries of definite parity:

$$(a^1 \otimes \cdots \otimes a^n)^\dagger = \omega^{\sum_{i<j} \sigma(a_i)\sigma(a_j)} a^{1\dagger} \otimes \cdots \otimes a^{n\dagger}.$$

We note that this involution corresponds to the result concerning adjoints of iterated quantum stochastic integrals given in proposition 9.7. Our first task in this section is to define a family of products in $T(\mathcal{I})$. The products will be denoted $^A\tilde{\circ}^B$ and are indexed over finite subsets of $\mathbf{N}$.

Let $A = \{i_1 < \cdots < i_{|A|}\}$ and $B = \{j_1 < \cdots < i_{|B|}\}$ be subsets of M_n so that $A \cup B = M_n$. We wish to define the product $^A\tilde{\circ}^B : \otimes^{|A|}\mathcal{I} \times \otimes^{|B|}\mathcal{I} \to \otimes^n\mathcal{I}$ where, as in Chapter 6, $\otimes^k\mathcal{I}$ denotes the tensor product $\mathcal{I} \otimes \cdots \otimes \mathcal{I}$ (k copies of $\mathcal{I}$). Let a and b be product tensors with entries of definite parity so that $a = a^1 \otimes \cdots \otimes a^{|A|}$ and $b = b^1 \otimes \cdots \otimes b^{|B|}$. If we write

$$a\, ^A\tilde{\circ}^B\, b = \omega^{\sum_{j_m < i_l} \sigma(a^l)\sigma(b^m)} c^1 \otimes \cdots \otimes c^n$$

then the product $^A\tilde{\circ}^B$ is defined fully if we set

$$c^k = \begin{cases} a^l & \text{if } k = i_l \in A \cap (M_n \setminus B); \\ b^m & \text{if } k = j_m \in (M_n \setminus A) \cap B; \\ a^l b^m & \text{if } k = i_l = j_m \in A \cap B. \end{cases}$$

We may now define the product $\tilde{*}$ in $T(\mathcal{I})$ component-wise by

$$(a\tilde{*}b)_m = \sum_{A \cup B = M_m} a_{|A|}\, ^A\tilde{\circ}^B\, b_{|B|}.$$

In the theorem that follows it will be seen that $\tilde{*}$ provides a means of multiplying together iterated quantum stochastic integrals.

Theorem 9.9 For arbitrary a and b in $T(\mathcal{I})$, arbitrary elements χ, ψ in $\mathcal{E}$ and arbitrary $t \geq 0$ we have that

$$\langle I(a)_t^\dagger \chi, I(b)_t \psi \rangle = \langle \chi, I(a\tilde{*}b)\psi \rangle.$$

Proof. This is entirely analogous to the proof of theorem 7.4 in Section 7.6 and is therefore omitted.∎

9.8 Commutation Relations

It is tempting to conjecture that the $\mathbf{Z}_n$-graded quantum stochastic calculus will yield graded commutation relations along the lines of those obtained for the $\mathbf{Z}_2$-graded case in Chapter 4. Let $\omega : \mathbf{Z}_n \times \mathbf{Z}_n \to \mathbf{C}$ be some map providing a generalised commutation factor. Using the differential form of the second fundamental formula given in Section 9.6 we see that the ω-commutator

$$d\left(\tilde{\Lambda}_A \tilde{\Lambda}_B - \omega(\sigma(A), \sigma(B))\tilde{\Lambda}_B \tilde{\Lambda}_A\right)$$

is equal to

$$\tilde{\Lambda}_A \otimes d\tilde{\Lambda}_B + \omega^{\sigma(A)\sigma(B)} \tilde{\Lambda}_B \otimes d\tilde{\Lambda}_A + 1 \otimes d\tilde{\Lambda}_{A.B}$$
$$- \omega(\sigma(A), \sigma(B))\tilde{\Lambda}_B \otimes d\tilde{\Lambda}_A - \omega^{\sigma(B)\sigma(A)}\omega(\sigma(A), \sigma(B))\tilde{\Lambda}_A \otimes d\tilde{\Lambda}_B$$
$$- \omega(\sigma(A), \sigma(B))1 \otimes d\tilde{\Lambda}_{B.A}.$$

In order for an interesting theory to be developed we require that the iterated integral terms cancel. For this to happen we require that

$$\omega(\sigma(A), \sigma(B)) = \omega^{\sigma(A)\sigma(B)}$$

in order for the $\tilde{\Lambda}_B \otimes d\tilde{\Lambda}_A$ terms to cancel. We also require that the $\tilde{\Lambda}_A \otimes d\tilde{\Lambda}_B$ terms cancel and for this to happen it is necessary that

$$\omega(\sigma(A), \sigma(B))\omega^{\sigma(A)\sigma(B)} = 1.$$

In order for both these conditions to be satisfied it is necessary that the only values taken by $\omega^{\sigma(A)\sigma(B)}$ and $\omega(\cdot, \cdot)$ are 1 and -1. This amounts to a requirement that the grading of the calculus in question must be a $\mathbf{Z}_2$ grading and so the tempting conjecture turns out to be false. In the next chapter it will be seen that attractive commutation relations may be obtained by using a grading over spaces of the form $\mathbf{Z}_n \times \mathbf{Z}_n$.

9.9 Alternative Sources of Commutation Relations

For all positive integers n, the group $\mathbf{Z}_n \times \mathbf{Z}_n$ has its composition defined by

$$(a, a') + (b, b') = (a + b, a' + b')$$

where a, a', b, b' are arbitrary elements of $\mathbf{Z}_n$.

In this section we show how $\mathbf{Z}_n \times \mathbf{Z}_n$ can be used to grade $n^2 - 1$ dimensional quantum stochastic calculus. First it is necessary to grade the indexing matrix algebra $M_0(n^2 - 1)$.

It suffices to define the grade of an arbitrary basis element E_β^α of $M_0(n^2 - 1)$. It is certainly true that we may write α uniquely in the form $nk + k'$ where k, k' are elements of $\mathbf{Z}_n$ and similarly we have that $\beta = nl + l'$ with l, l' in $\mathbf{Z}_n$. Given these decompositions we assign the grade $(k - l, k' - l')$ to E_β^α where the subtraction is in the $\mathbf{Z}_n$ sense. As might be expected, we define $\sigma(E_\beta^\alpha)$ to be $(k - l, k' - l')$. With this notation we may define the subspaces $M_0(n^2 - 1)_\gamma$ of $M_0(n^2 - 1)$ with $\gamma \in \mathbf{Z}_n \times \mathbf{Z}_n$ by

$$M_0(n^2 - 1)_{(a,b)} = \text{span}\left\{ E_\beta^\alpha : \sigma(E_\beta^\alpha) = (a, b) \right\}.$$

It is clear that $M_0(n^2 - 1)$ is equal to the internal direct sum

$$\sum_{\gamma \in \mathbf{Z}_n \times \mathbf{Z}_n} M_0(n^2 - 1)_\gamma$$

so we have a $\mathbf{Z}_n \times \mathbf{Z}_n$-graded vector space structure for $M_0(n^2 - 1)$. Now observe that, for $i = nk + k' \neq 0$, $\alpha = nl + l'$ and $\beta = nm + m'$ with $k, l, m,$ k', l' and m' all in $\mathbf{Z}_n$ we have

$$\sigma(E_\beta^i) + \sigma(E_i^\alpha) = (k - m, k' - m') + (l - k, l' - k') = (l - m, l' - m') = \sigma(E_\beta^\alpha)$$

so that, for all $\alpha, \beta, \gamma, \delta$ with $0 \leq \alpha, \beta, \gamma, \delta \leq n^2 - 1$ we have

$$\sigma(E_\beta^\alpha . E_\delta^\gamma) = \sigma(E_\beta^\alpha) + \sigma(E_\delta^\gamma).$$

Thus the graded algebra condition

$$M_0(n^2 - 1)_\gamma . M_0(n^2 - 1)_{\gamma'} \subset M_0(n^2 - 1)_{\gamma + \gamma'}$$

given in (9.1) holds for all γ and γ' in $\mathbf{Z_n} \times \mathbf{Z_n}$ and we have a $\mathbf{Z_n} \times \mathbf{Z_n}$-graded algebra structure for $M_0(n^2 - 1)$.

Let z be fixed as an arbitrary n^{th} root of 1 and the function $\omega : \mathbf{Z}_n \times \mathbf{Z}_n \to \mathbf{C}$ be defined by

$$\omega((a_1, a_2), (b_1, b_2)) = z^{a_1 b_2 - a_2 b_1}.$$

A map $\alpha : G \times G \to \{w \in \mathbf{C} : |w| = 1\}$ where G is an abelian group is said to be a *bicharacter* if, for all a, b and c in G, we have that

$$\alpha(b, a)\alpha(c, a + b) = \alpha(b + c, a)\alpha(c, b)$$

and

$$\alpha(a, 0) = \alpha(0, a) = 1.$$

We will now show that the map ω defined above is a bicharacter. Let $(a_1, a_2), (b_1, b_2), (c_1, c_2)$ be arbitrary elements of $\mathbf{Z}_n \times \mathbf{Z}_n$. Then we have that

$$\omega((b_1, b_2), (a_1, a_2))\omega((c_1, c_2), (a_1, a_2) + (b_1, b_2))$$
$$=\omega((b_1, b_2), (a_1, a_2))\omega((c_1, c_2), (a_1 + b_1, a_2 + b_2))$$
$$=z^{b_1 a_2 - b_2 a_1} z^{c_1 a_2 + c_1 b_2 - c_2 a_1 - c_2 b_1}$$
$$=z^{(b_1 + c_1)a_2 - (b_2 + c_2)a_1} z^{c_1 b_2 - c_2 b_1}$$
$$=\omega((b_1 + c_1, b_2 + c_2), (a_1, a_2))\omega((c_1, c_2), (b_1, b_2))$$
$$=\omega((b_1, b_2) + (c_1, c_2), (a_1, a_2))\omega((c_1, c_2), (b_1, b_2))$$

so that the first condition is satisfied. For the second condition we have for arbitrary (a_1, a_2) in $\mathbf{Z}_n \times \mathbf{Z}_n$ that

$$\omega((a_1, a_2), 0) = \omega((a_1, a_2), (0, 0)) = z^{a_1 0 - a_2 0} = 1$$

and

$$\omega(0, (a_1, a_2)) = \omega((0, 0), (a_1, a_2)) = z^{0 a_1 - 0 a_2} = 1$$

as required. Thus ω is a bicharacter.

A bicharacter such as ω may be used as a commutation factor in the formation of tensor products of G-graded algebras. Such a tensor product is written $A \otimes_\omega B$ and a pair of product tensors in $A \otimes_\omega B$ with entries of definite parity are multiplied together according to the following rule:

$$(a_1 \otimes a_2)(b_1 \otimes b_2) = \omega(\sigma(b_1), \sigma(a_2))a_1 b_1 \otimes a_2 b_2.$$

The graded theories of quantum stochastic calculus described in this work may all be formulated using tensor products of this type in the formation of the quantum stochastic integrals from elementary processes. The $\mathbf{Z}_2$-graded calculus is obtained by using the bicharacter $(a, b) \mapsto (-1)^{ab}$ where a, b are elements of $\mathbf{Z}_2$. The $\mathbf{Z}_n$-graded theory described earlier in this chapter is obtained by using the bicharacter which maps a pair of elements (a, b) in $\mathbf{Z}_n \times \mathbf{Z}_n$ to ω^{ab} where here ω indicates a fixed primitive n^{th} root of 1.

While such a formulation of graded quantum stochastic calculus in the $\mathbf{Z}_2$-graded and $\mathbf{Z}_n$-graded case is simply an alternative to methods that have already been established, the usage of tensor products endowed with a bicharacter commutation factor is natural when grading over a general abelian group. Given an element A of definite grade in the $\mathbf{Z}_n \times \mathbf{Z}_n$-graded version of $M_0(n^2 - 1)$ described above, the grade that will be assigned to the quantum stochastic integrator $d\tilde{\Lambda}_A$ is the grade of A. The process $\tilde{\Lambda}_A$ is defined in the obvious way as

$$\int_0^{\cdot} Id \otimes_\omega d\tilde{\Lambda}_A.$$

Details of this will not be given. By applying the differential form of the second fundamental formula to the formal graded commutator $d(\tilde{\Lambda}_A \tilde{\Lambda}_B - \omega(\sigma(B), \sigma(A))\tilde{\Lambda}_B \tilde{\Lambda}_A)$ with homogeneous A, B we obtain

$$\Lambda_A \otimes_\omega d\Lambda_B + \omega(\sigma(B), \sigma(A))\Lambda_B \otimes_\omega d\Lambda_A + Id \otimes_\omega d\Lambda_{A((\hat{\delta}))B}$$
$$-\omega(\sigma(B), \sigma(A))\Lambda_B \otimes_\omega d\Lambda_A - \omega(\sigma(B), \sigma(A))\omega(\sigma(A), \sigma(B))\Lambda_A \otimes_\omega d\Lambda_B$$
$$-\omega(\sigma(B), \sigma(A))Id \otimes_\omega d\Lambda_{B((\hat{\delta}))A}.$$

It is clear that the second and fourth terms cancel. The coefficient taken by the fifth term is equal to $z^{a_1 b_2 - b_1 a_2} z^{b_1 a_2 - a_1 b_2}$ where (a_1, a_2) is the grade of A and (b_1, b_2) is the grade of B in the $\mathbf{Z}_n \times \mathbf{Z}_n$-graded $M_0(n^2 - 1)$. This coefficient is clearly equal to 1 as $z \neq 0$ and so the fifth term cancels with the first term and we are left with

$$Id \otimes_\omega d\Lambda_{A((\hat{\delta}))B} - \omega(\sigma(B), \sigma(A))Id \otimes_\omega d\Lambda_{B((\hat{\delta}))A}.$$

Let the formal graded commutator of $\mathbf{Z}_n \times \mathbf{Z}_n$-graded quantum stochastic process be defined by linear extension of the following rule for homogeneous processes E, F:

$$EF - \omega(\sigma(F), \sigma(E))FE.$$

We denote this commutator by $[\cdot, \cdot]_\omega$. Similarly, the graded commutator defined by linear extension of the following rule for arbitrary homogeneous A, B in the $\mathbf{Z}_n \times \mathbf{Z}_n$-graded $M_0(n^2 - 1)$:

$$A.B - \omega(\sigma(B), \sigma(A))B.A$$

is also denoted $[\cdot, \cdot]_\omega$. With this notation and the results of the previous paragraph we see that for all A, B in $M_0(n^2 - 1)$ we have

$$[\tilde{\Lambda}_A, \tilde{\Lambda}_B]_\omega = \tilde{\Lambda}_{[A,B]_\omega}. \tag{9.5}$$

Thus we have obtained a commutation result similar to the one presented in Chapter 4. It is clear that an arbitrary finite abelian group G may be used to grade N-dimensional quantum stochastic calculus provided that G can be made to grade $M_0(N)$ and a bicharacter α can be defined on G. Furthermore, it is clear that graded commutation relations of the form (9.5) will be obtained if the bicharacter α satisfies the property

$$\alpha(g_1, g_2)\alpha(g_2, g_1) = 1$$

for all g_1, g_2 in G.

9.10 Z_n-Graded Quantum Stochastic Calculus With Infinite Degrees of Freedom

The work done so far has introduced a number of different gradings for N-dimensional quantum stochastic calculus, this being associated with Fock space over $L^2(\mathbf{R}_+; \mathbf{C}^N)$. In [P1] it is shown how, with a few caveats, a quantum stochastic calculus taking an infinite number of degrees of freedom may be constructed in Fock space over $L^2(\mathbf{R}_+; \mathcal{V})$ where $\mathcal{V}$ is an arbitrary countably infinite dimensional Hilbert space. In this section it is shown how a Z_n-grading may be applied to such a calculus. As the theory develops it will become clear by what means the Z_n-grading in the N-dimensional case might have been performed (profitlessly) using arbitrary partitions of M_N rather than the inequalities actually used.

We begin with a result given in Section 27 of [P1]. Let $(L_\alpha^\beta)_{\alpha,\beta\in\mathbf{N}}$ be a family of quantum stochastic processes. For each n in $\mathbf{N}$ define $X_n(t)$ by

$$X_n(t) = \sum_{0\leq\alpha,\beta\leq n} \int_0^t L_\alpha^\beta(s)\,d\Lambda_\beta^\alpha(s).$$

Define $\mathcal{M}$ to be the set

$$\{f \in \mathbf{L}^2(\mathbf{R}_+; \mathcal{V}) : f^i = 0 \text{ for all but a finite number of indices } i\}.$$

Note that $\mathcal{M}$ is dense in $L^2(\mathbf{R}_+; \mathcal{V})$. Proposition 27.1 of [P1] states that, if for all $t \geq 0$, for all $f \in \mathcal{M}$ and for all $\alpha \geq 0$ the L_α^β have the property

$$\int_0^t \sum_{\beta=0}^\infty \|L_\alpha^\beta(s)e(f)\|^2(1 + \|f(s)\|^2)\,ds < \infty, \tag{9.6}$$

then there exists a quantum stochastic process X such that for each f in $\mathcal{M}$ and each $t \geq 0$

$$\lim_{n\to\infty} \sup_{0\leq s\leq t} \|(X_n(s) - X(s))e(f)\| = 0.$$

Seeing as our purpose is simply to grade the integrators of this theory, it suffices to restrict our attention to processes L_α^β of the form $\lambda_\alpha^\beta Id$ where λ_α^β is an element of $\mathbf{C}$. With this simplification, condition (9.6) becomes

$$\int_0^t \sum_{\beta=0}^\infty \|\lambda_\alpha^\beta e(f)\|^2(1 + \|f(s)\|^2)\,ds < \infty.$$

This amounts to a condition that, for each $\alpha \geq 0$

$$\sum_{\beta=0}^\infty |\lambda_\alpha^\beta|^2 < \infty. \tag{9.7}$$

In other words we require that for each $\alpha \geq 0$ the sequence $(\lambda_\alpha^\beta)_{\alpha \in \mathbb{N}}$ is an $l^2(\mathbf{C})$ sequence. Therefore, given such a family of sequences, the quantum stochastic process

$$\int_0^{\cdot} \lambda_\alpha^\beta Id \otimes d\Lambda_\beta^\alpha(s)$$

is well-defined.

While condition (9.7) provides the greatest variety of permissible integrators, it is not of practical value in this work. In order to apply the methods already developed we require an Ito *algebra*. The following example shows that the Ito product of two integrators of the form described above is not necessarily itself a permissible integrator.

Define the families of complex numbers $(\lambda_\alpha^\beta)_{\alpha,\beta \in \mathbb{N}}$, $(\mu_\gamma^\delta)_{\gamma,\delta \in \mathbb{N}}$ by

$$\lambda_\alpha^\beta = \begin{cases} \alpha!, & \text{if } \beta = 1, \ \alpha > 0; \\ 0, & \text{otherwise.} \end{cases}$$

$$\mu_\gamma^\delta = \begin{cases} (\delta!)^{-1}, & \text{if } \gamma = 1, \ \delta > 0; \\ 0, & \text{otherwise.} \end{cases}$$

The integrators $\lambda_\alpha^\beta \, d\Lambda_\beta^\alpha$ and $\mu_\gamma^\delta \, d\Lambda_\gamma^\delta$ both give well-defined integrals according to the conditions described above. However, the Ito product

$$\lambda_\alpha^\beta \, d\Lambda_\beta^\alpha . \mu_\gamma^\delta \, d\Lambda_\delta^\gamma$$

would be equal to $\lambda_i^1 \mu_1^i \, d\Lambda_1^1$ if this expression did not involve the divergent sum $\sum_{i=1}^{\infty} 1$ which means that the product is undefined. Similar problems are encountered when taking adjoints.

We see that the conditions for permissible integrators must be tightened if a useful theory is to be developed. The obvious means of doing this is to require that the family of coefficients $(\lambda_\alpha^\beta)_{\alpha,\beta \in \mathbb{N}}$ satisfies the condition

$$\sum_{\alpha,\beta=0}^{\infty} |\lambda_\alpha^\beta|^2 < \infty. \tag{9.8}$$

It is certainly true that if the λ_α^β satisfy this condition, condition (9.7) is satisfied and so an integrator of the form $\lambda_\alpha^\beta d\Lambda_\beta^\alpha$ is well-defined. If the family of complex numbers $(\mu_\gamma^\delta)_{\gamma,\delta \in \mathbb{N}}$ also satisfies (9.8) then the Ito product

$$\lambda_\alpha^\beta \, d\Lambda_\beta^\alpha . \mu_\gamma^\delta \, d\Lambda_\delta^\gamma$$

is equal to $\lambda_i^\beta \mu_\gamma^i \, d\Lambda_\beta^\gamma$ which is well defined because, making use of the Cauchy-Schwarz inequality, we can see that

$$\sum_{\beta,\gamma=0}^{\infty} |\lambda_i^\beta \mu_\gamma^i|^2 \leq \sum_{\beta,\gamma=0}^{\infty} \left(\sum_{i=0}^{\infty} |\lambda_i^\beta|^2 \right) \left(\sum_{i=0}^{\infty} |\mu_\gamma^i|^2 \right).$$

This is clearly equal to

$$\left(\sum_{\beta=0}^{\infty}\sum_{i=1}^{\infty}|\lambda_i^{\beta}|^2\right)\left(\sum_{\gamma=0}^{\infty}\sum_{i=1}^{\infty}|\mu_{\gamma}^i|^2\right)$$

which is finite by condition (9.8).

Of course, such arrays of complex numbers $(\lambda_{\alpha}^{\beta})_{\alpha,\beta\in\mathbf{N}}$ as those satisfying (9.8) can be interpreted as bounded linear transformations from $l^2(\mathbf{C})$ to $l^2(\mathbf{C})$. If e_{γ} denotes the element of $l^2(\mathbf{C})$ which has zero everywhere except for the γth entry which is 1 then the array $((\lambda_{\alpha}^{\beta}))$ may be interpreted as a linear transformation by

$$((\lambda_{\alpha}^{\beta}))e_{\gamma} = (\lambda_{\gamma}^{\beta})_{\beta\geq 0}.$$

Thus we shall index this infinite-dimensional quantum stochastic calculus using the space $M_0(\mathbf{N})$ of all bounded linear operators in $l^2(\mathbf{C})$. Let E_{β}^{α} denote the transformation which maps e_{α} to e_{β} and sends all e_{γ} for which $\gamma \neq \alpha$ to 0. Any element of $M_0(\mathbf{N})$ can be expressed as a (not necessarily finite) complex linear combination of the E_{β}^{α}. As might be expected, the multiplication in $M_0(\mathbf{N})$ is defined by linear extension of the rule

$$E_{\beta}^{\alpha}.E_{\delta}^{\gamma} = \hat{\delta}_{\delta}^{\alpha} E_{\beta}^{\gamma}.$$

We will now construct a $\mathbf{Z}_n$-grading for $M_0(\mathbf{N})$. It is clear that the approach used in Section 9.2 of this chapter will not work in this context. This is easily seen from the fact that inequalities of the kind used in the finite dimensional case will not permit a finite dimensional subspace of $M_0(\mathbf{N})$ to be assigned grade $n-1$. Instead we partition $\mathbf{N}$ into n sets labelled $A_0,\ldots,A_{n-1}$. The partition condition means that we have $\cup_{\alpha=0}^{n-1}A_{\alpha} = \mathbf{N}$ and that, for all $0 \leq \alpha \neq \beta \leq n-1$, the intersection $A_{\alpha} \cap A_{\beta}$ is the null set $\emptyset$. It is necessary to require that $0 \in A_0$. This partition will be considered to be fixed throughout the remainder of this section.

We may now grade $M_0(\mathbf{N})$ by means of assigning grades to the elements E_{β}^{α}. If $\alpha \in A_{\gamma_1}$ and $\beta \in A_{\gamma_2}$ then the element E_{β}^{α} is assigned grade $\gamma_1 - \gamma_2$. As in the N-dimensional case we denote the grade of E_{β}^{α} by σ_{β}^{α}. As before, we have that $\sigma_{\beta}^{\alpha} = \sigma_0^{\alpha} - \sigma_0^{\beta}$, $\sigma_{\alpha}^{\beta} = -\sigma_{\alpha}^{\beta}$ and $\sigma(E_{\alpha}^{\alpha}.E_{\delta}^{\gamma}) = \sigma_{\beta}^{\alpha} + \sigma_{\delta}^{\gamma}$. The graded $M_0(\mathbf{N})$ will be denoted $M_0(\mathbf{N},\mathbf{A})$. Equipped with $M_0(\mathbf{N},\mathbf{A})$ we now produce a $\mathbf{Z}_n$-graded version of infinite dimensional quantum stochastic calculus. As in Section 9.2, let ω be a primitive n^{th} root of 1. Define a map B in $L^2(\mathbf{R}_+;\mathcal{V})$ component-wise by

$$(Bf)^i = \omega^{\sigma(E_0^i)}f^i$$

where $f = (f^1, f^2,\ldots)$ is an arbitrary element of $\mathcal{M}$.

The grading process G may now be defined on $\mathcal{E}$. In this context, $\mathcal{E}$ consists of all finite complex linear combinations of exponential vectors $e(f)$ with $f \in \mathcal{M}$. Note that, by proposition 2.4, the density of $\mathcal{M}$ in $L^2(\mathbf{R}_+,\mathcal{V})$ means that $\mathcal{E}$ is dense in $\Gamma(L^2(\mathbf{R}_+; \mathcal{V}))$. We define G by linear extension of the following rule for an arbitrary exponential vector $e(f)$ with $f \in \mathcal{M}$ and an arbitrary time $t \geq 0$:

$$G(t)e(f) = e(\chi_{[0,t]}Bf + \chi_{(t,\infty)}).$$

It is easy to see from the unitarity of each $G(t)$ that $G^{n-1} = G^{-1} = G^\dagger$.

The differential process $d\tilde{\Lambda}_\beta^\alpha$ is defined to be

$$G^{\sigma_\alpha^\beta} d\Lambda_\beta^\alpha$$

with the corresponding process $\tilde{\Lambda}_\beta^\alpha$ defined in the expected way as

$$\int_0^\cdot G^{\sigma_\alpha^\beta}(s) \otimes d\Lambda_\beta^\alpha(s).$$

For an arbitrary $C \in M_0(\mathbf{N}, \mathbf{A})$ with $C = \lambda_\alpha^\beta E_\beta^\alpha$ we define the process $\tilde{\Lambda}_C$ in the obvious way by
$$\tilde{\Lambda}_C = \lambda_\alpha^\beta \tilde{\Lambda}_\beta^\alpha.$$

It must be shown that this process exists according to condition (9.6). For $\alpha \geq 0$, arbitrary $f \in \mathcal{M}$ and arbitrary $t \geq 0$ we have that

$$\int_0^t \sum_{\beta=0}^\infty \|\lambda_\alpha^\beta G^{\sigma_\alpha^\beta}(s)e(f)\|(1 + \|f(s)\|^2)\, ds$$

is finite by the unitarity of G. We will conclude this section by showing that conjugation of the $\tilde{\Lambda}_\beta^\alpha$ by G yields the required grading factor.

Take arbitrary elements f, g of $\mathcal{M}$, arbitrary α, β in $\mathbf{N}$ and arbitrary $t \geq 0$. Then we may re-write the inner product $\langle e(f), G^{-1}(t)\tilde{\Lambda}_\beta^\alpha(t)G(t)e(g)\rangle$ as

$$\left\langle G(t)e(f), \int_0^t Id \otimes d\tilde{\Lambda}_\beta^\alpha(t)G(t)e(g)\right\rangle.$$

By the first fundamental formula stated in theorem 2.7 and the obvious extension of lemma 4.3 this is equal to

$$\int_0^t f_\beta(s)g^\alpha(s)\omega^{\sigma_\alpha^\beta}\langle e(f), G^{\sigma_\alpha^\beta}(s)e(g)\rangle\, ds$$

which, again by the first fundamental formula, is equal to

$$\langle e(f), \omega^{\sigma_\beta^\alpha} \tilde{\Lambda}_\beta^\alpha(t)e(g)\rangle.$$

Thus we have that the conjugate $G^{-1}\tilde{\Lambda}_\beta^\alpha G$ of a process of the form $\tilde{\Lambda}_\beta^\alpha$ in infinite dimensional $\mathbf{Z}_n$-graded quantum stochastic calculus is equal to $\omega^{\sigma_\beta^\alpha}\tilde{\Lambda}_\beta^\alpha$ which is what would be expected.

When generalising this result to arbitrary homogeneous elements of $M_0(\mathbf{N}, \mathbf{A})$, concerns of convergence do not appear because the restriction to $\mathcal{M}$ ensures that all sums involved are finite.

References

[AH] D. B. Applebaum, R. L. Hudson: Fermion Ito's Formula and Stochastic Evolutions. Commun. Math. Phys. **96**, 473-496 (1984)

[B] N. Bourbaki: Algebra I, Chapters 1-3 (English translation). Addison-Wesley 1973

[BSW] C. Barnett, R. F. Streater, I. F. Wilde: The Ito-Clifford Integral. J. Funct. Anal. **48**, 172-212 (1982)

[C] C. Chevalley: The Construction And Study Of Certain Important Algebras. Publ. Math. Soc. Japan I, Princeton University Press, Princeton NJ, 1955

[CEH] P. B. Cohen, T. M. W. Eyre, R. L. Hudson: Higher Order Ito Product Formula and Generators of Evolutions and Flows. International Journal Of Theoretical Physics **34**, No. 8 1481-1486 (1995)

[CP] V. Chari, A. Pressley: A Guide to Quantum Groups. Cambridge University Press, Cambridge, 1994

[D] J. Dixmier: Algèbres Enveloppantes. Gauthier-Villars, Paris, 1974

[E] T. M. W. Eyre: Chaotic Expansions of Elements of the Universal Enveloping Superalgebra Associated with a Z_2-Graded Quantum Stochastic Calculus. Commun. Math. Phys.,**192** 1, 9-28 (1998)

[EH] T. M. W. Eyre, R. L. Hudson: Representations of Lie Superalgebras and Generalized Boson-Fermion Equivalence in Quantum Stochastic Calculus. Commun. Math. Phys. **186**, 87-94 (1997)

[Ev] M. P. Evans: Existence of Quantum Diffusions. Probability Theory and Related Fields **81**, 473-483 (1989)

[H2] R. L. Hudson: Lie Algebras and Chaotic Expansions in Quantum Stochastic Calculus. Proceedings, Gregynog, ed. I. Davies et al (1995)

[HP1] R. L. Hudson, K. R. Parthasarathy: Quantum Ito's Formula and Stochastic Evolutions. Commun. Math. Phys. **93**, 301-323 (1984)

[HP2] R. L. Hudson, K. R. Parthasarathy: Unification of Fermion and Boson Stochastic Calculus. Commun. Math. Phys. **104**, 457-470 (1986)

[HP3] R. L. Hudson, K. R. Parthasarathy: Casimir Chaos in a Boson Fock Space. J. Funct. Anal. **119**, 319-339 (1994)

[HP4] R. L. Hudson, K. R. Parthasarathy: The Casimir Chaos Map for $U(N)$. Tatra Mountains Math. Proc. **3**, 1-9 (1993)

[HP5] R. L. Hudson, K. R. Parthasarathy: Chaos Map for the Universal Enveloping Algebra of $U(N)$. Math. Proc. Camb. Phil. Soc. **117**, 21-30 (1995)

[HPu] R. L. Hudson, S. Pulmannova: Chaotic Expansion of Elements of the Universal Enveloping Algebra of a Lie Algebra Associated with a Quantum Stochastic Calculus. Proc. LMS, *to appear*

[HS] R. L. Hudson, V. R. Struleckaja: Fermionic Quantum Stochastic Flows. Letters in Mathematical Physics **37**, 309-317 (1996)

[Hum] J. E. Humphreys: Introduction to Lie Algebras and Representation Theory. Graduate Texts in Mathematics Vol. **9**, Springer, New York, 1972

[J]   N. Jacobson: Lie Algebras. Dover, New York, 1979

[K]   V. G. Kac, Lie Superalgebras. Advances In Mathematics **26**, 8-96 (1977)

[L]   J. M. Lindsay: Independence For Quantum Stochastic Integrators. Quantum Probability and Related Topics Vol. **VI**, 325-332 (1991)

[P1]   K. R. Parthasarathy: An Introduction to Quantum Stochastic Calculus. Birkhäuser, Basel, 1992

[P2]   K. R. Parthasarathy: Some New Examples of Lie Super–Algebra Representations Arising From Quantum Stochastic Calculus. Infinite Dimensional Analysis and Quantum Probability, Vol. 1 No. 1, 33-42

[P3]   K. R. Parthasarathy: Quantum Stochastic Calculus. Proceedings of the International Congress of Mathematicians, Zürich, Switzerland 1994. Birkhäuser Verlag, Basel (1995)

[S]   M. Scheunert: The Theory of Lie Superalgebras. Lecture Notes in Mathematics Vol. **716**, Springer, Berlin, 1979

[Sch1]   M. Schürmann: Noncommutative Stochastic Processes with Independent and Stationary Increments Satisfy Quantum Stochastic Differential Equations. Probability Theory and Related Fields **84**, 473-490 (1990)

[Sch2]   M. Schürman: White Noise on Bialgebras. Lecture Notes in Mathematics Vol. **1544**, Springer, Berlin, 1993

Index

Springer
and the
environment

At Springer we firmly believe that an international science publisher has a special obligation to the environment, and our corporate policies consistently reflect this conviction.
We also expect our business partners – paper mills, printers, packaging manufacturers, etc. – to commit themselves to using materials and production processes that do not harm the environment. The paper in this book is made from low- or no-chlorine pulp and is acid free, in conformance with international standards for paper permanency.

 Springer

Lecture Notes in Mathematics

For information about Vols. 1–1504
please contact your bookseller or Springer-Verlag

Vol. 1596: L. Heindorf, L. B. Shapiro, Nearly Projective Boolean Algebras. X, 202 pages. 1994.

Vol. 1597: B. Herzog, Kodaira-Spencer Maps in Local Algebra. XVII, 176 pages. 1994.

Vol. 1598: J. Berndt, F. Tricerri, L. Vanhecke, Generalized Heisenberg Groups and Damek-Ricci Harmonic Spaces. VIII, 125 pages. 1995.

Vol. 1599: K. Johannson, Topology and Combinatorics of 3-Manifolds. XVIII, 446 pages. 1995.

Vol. 1600: W. Narkiewicz, Polynomial Mappings. VII, 130 pages. 1995.

Vol. 1601: A. Pott, Finite Geometry and Character Theory. VII, 181 pages. 1995.

Vol. 1602: J. Winkelmann, The Classification of Three-dimensional Homogeneous Complex Manifolds. XI, 230 pages. 1995.

Vol. 1603: V. Ene, Real Functions – Current Topics. XIII, 310 pages. 1995.

Vol. 1604: A. Huber, Mixed Motives and their Realization in Derived Categories. XV, 207 pages. 1995.

Vol. 1605: L. B. Wahlbin, Superconvergence in Galerkin Finite Element Methods. XI, 166 pages. 1995.

Vol. 1606: P.-D. Liu, M. Qian, Smooth Ergodic Theory of Random Dynamical Systems. XI, 221 pages. 1995.

Vol. 1607: G. Schwarz, Hodge Decomposition – A Method for Solving Boundary Value Problems. VII, 155 pages. 1995.

Vol. 1608: P. Biane, R. Durrett, Lectures on Probability Theory. Editor: P. Bernard. VII, 210 pages. 1995.

Vol. 1609: L. Arnold, C. Jones, K. Mischaikow, G. Raugel, Dynamical Systems. Montecatini Terme, 1994. Editor: R. Johnson. VIII, 329 pages. 1995.

Vol. 1610: A. S. Üstünel, An Introduction to Analysis on Wiener Space. X, 95 pages. 1995.

Vol. 1611: N. Knarr, Translation Planes. VI, 112 pages. 1995.

Vol. 1612: W. Kühnel, Tight Polyhedral Submanifolds and Tight Triangulations. VII, 122 pages. 1995.

Vol. 1613: J. Azéma, M. Emery, P. A. Meyer, M. Yor (Eds.), Séminaire de Probabilités XXIX. VI, 326 pages. 1995.

Vol. 1614: A. Koshelev, Regularity Problem for Quasilinear Elliptic and Parabolic Systems. XXI, 255 pages. 1995.

Vol. 1615: D. B. Massey, Le Cycles and Hypersurface Singularities. XI, 131 pages. 1995.

Vol. 1616: I. Moerdijk, Classifying Spaces and Classifying Topoi. VII, 94 pages. 1995.

Vol. 1617: V. Yurinsky, Sums and Gaussian Vectors. XI, 305 pages. 1995.

Vol. 1618: G. Pisier, Similarity Problems and Completely Bounded Maps. VII, 156 pages. 1996.

Vol. 1619: E. Landvogt, A Compactification of the Bruhat-Tits Building. VII, 152 pages. 1996.

Vol. 1620: R. Donagi, B. Dubrovin, E. Frenkel, E. Previato, Integrable Systems and Quantum Groups. Montecatini Terme, 1993. Editors:M. Francaviglia, S. Greco. VIII, 488 pages. 1996.

Vol. 1621: H. Bass, M. V. Otero-Espinar, D. N. Rockmore, C. P. L. Tresser, Cyclic Renormalization and Auto-morphism Groups of Rooted Trees. XXI, 136 pages. 1996.

Vol. 1622: E. D. Farjoun, Cellular Spaces, Null Spaces and Homotopy Localization. XIV, 199 pages. 1996.

Vol. 1623: H.P. Yap, Total Colourings of Graphs. VIII, 131 pages. 1996.

Vol. 1624: V. Brînzanescu, Holomorphic Vector Bundles over Compact Complex Surfaces. X, 170 pages. 1996.

Vol.1625: S. Lang, Topics in Cohomology of Groups. VII, 226 pages. 1996.

Vol. 1626: J. Azéma, M. Emery. M. Yor (Eds.), Séminaire de Probabilités XXX. VIII, 382 pages. 1996.

Vol. 1627: C. Graham, Th. G. Kurtz, S. Méléard, Ph. E. Protter, M. Pulvirenti, D. Talay, Probabilistic Models for Nonlinear Partial Differential Equations. Montecatini Terme, 1995. Editors: D. Talay, L. Tubaro. X, 301 pages. 1996.

Vol. 1628: P.-H. Zieschang, An Algebraic Approach to Association Schemes. XII, 189 pages. 1996.

Vol. 1629: J. D. Moore, Lectures on Seiberg-Witten Invariants. VII, 105 pages. 1996.

Vol. 1630: D. Neuenschwander, Probabilities on the Heisenberg Group: Limit Theorems and Brownian Motion. VIII, 139 pages. 1996.

Vol. 1631: K. Nishioka, Mahler Functions and Transcendence. VIII, 185 pages.1996.

Vol. 1632: A. Kushkuley, Z. Balanov, Geometric Methods in Degree Theory for Equivariant Maps. VII, 136 pages. 1996.

Vol.1633: H. Aikawa, M. Essén, Potential Theory – Selected Topics. IX, 200 pages.1996.

Vol. 1634: J. Xu, Flat Covers of Modules. IX, 161 pages. 1996.

Vol. 1635: E. Hebey, Sobolev Spaces on Riemannian Manifolds. X, 116 pages. 1996.

Vol. 1636: M. A. Marshall, Spaces of Orderings and Abstract Real Spectra. VI, 190 pages. 1996.

Vol. 1637: B. Hunt, The Geometry of some special Arithmetic Quotients. XIII, 332 pages. 1996.

Vol. 1638: P. Vanhaecke, Integrable Systems in the realm of Algebraic Geometry. VIII, 218 pages. 1996.

Vol. 1639: K. Dekimpe, Almost-Bieberbach Groups: Affine and Polynomial Structures. X, 259 pages. 1996.

Vol. 1640: G. Boillat, C. M. Dafermos. P. D. Lax, T. P. Liu, Recent Mathematical Methods in Nonlinear Wave Propagation. Montecatini Terme, 1994. Editor: T. Ruggeri. VII, 142 pages. 1996.

Vol. 1641: P. Abramenko, Twin Buildings and Applications to S-Arithmetic Groups. IX, 123 pages. 1996.

Vol. 1642: M. Puschnigg, Asymptotic Cyclic Cohomology. XXII, 138 pages. 1996.

Vol. 1643: J. Richter-Gebert, Realization Spaces of Polytopes. XI, 187 pages. 1996.

Vol. 1644: A. Adler, S. Ramanan. Moduli of Abelian Varieties. VI, 196 pages. 1996.

Vol. 1645: H. W. Broer, G. B. Huitema, M. B. Sevryuk, Quasi-Periodic Motions in Families of Dynamical Systems. XI, 195 pages. 1996.

Vol. 1646: J.-P. Demailly, T. Peternell, G. Tian, A. N. Tyurin, Transcendental Methods in Algebraic Geometry. Cetraro, 1994. Editors: F. Catanese, C. Ciliberto. VII, 257 pages. 1996.

Vol. 1647: D. Dias, P. Le Barz, Configuration Spaces over Hilbert Schemes and Applications. VII. 143 pages. 1996.

Vol. 1648: R. Dobrushin, P. Groeneboom, M. Ledoux, Lectures on Probability Theory and Statistics. Editor: P. Bernard. VIII, 300 pages. 1996.

Vol. 1649: S. Kumar, G. Laumon, U. Stuhler, Vector Bundles on Curves – New Directions. Cetraro, 1995. Editor: M. S. Narasimhan. VII, 193 pages. 1997.

Vol. 1650: J. Wildeshaus, Realizations of Polylogarithms. XI, 343 pages. 1997.

Vol. 1651: M. Drmota, R. F. Tichy, Sequences, Discrepancies and Applications. XIII, 503 pages. 1997.

Vol. 1652: S. Todorcevic, Topics in Topology. VIII, 153 pages. 1997.

Vol. 1653: R. Benedetti, C. Petronio, Branched Standard Spines of 3-manifolds. VIII, 132 pages. 1997.

Vol. 1654: R. W. Ghrist, P. J. Holmes, M. C. Sullivan, Knots and Links in Three-Dimensional Flows. X, 208 pages. 1997.

Vol. 1655: J. Azéma, M. Emery, M. Yor (Eds.), Séminaire de Probabilités XXXI. VIII, 329 pages. 1997.

Vol. 1656: B. Biais, T. Björk, J. Cvitanic, N. El Karoui, E. Jouini, J. C. Rochet, Financial Mathematics. Bressanone, 1996. Editor: W. J. Runggaldier. VII, 316 pages. 1997.

Vol. 1657: H. Reimann, The semi-simple zeta function of quaternionic Shimura varieties. IX, 143 pages. 1997.

Vol. 1658: A. Pumarino, J. A. Rodríguez, Coexistence and Persistence of Strange Attractors. VIII, 195 pages. 1997.

Vol. 1659: V, Kozlov, V. Maz'ya, Theory of a Higher-Order Sturm-Liouville Equation. XI, 140 pages. 1997.

Vol. 1660: M. Bardi, M. G. Crandall, L. C. Evans, H. M. Soner, P. E. Souganidis, Viscosity Solutions and Applications. Montecatini Terme, 1995. Editors: I. Capuzzo Dolcetta, P. L. Lions. IX, 259 pages. 1997.

Vol. 1661: A. Tralle, J. Oprea, Symplectic Manifolds with no Kähler Structure. VIII, 207 pages. 1997.

Vol. 1662: J. W. Rutter, Spaces of Homotopy Self-Equivalences – A Survey. IX, 170 pages. 1997.

Vol. 1663: Y. E. Karpeshina; Perturbation Theory for the Schrödinger Operator with a Periodic Potential. VII, 352 pages. 1997.

Vol. 1664: M. Väth, Ideal Spaces. V, 146 pages. 1997.

Vol. 1665: E. Giné, G. R. Grimmett, L. Saloff-Coste, Lectures on Probability Theory and Statistics 1996. Editor: P. Bernard. X, 424 pages. 1997.

Vol. 1666: M. van der Put, M. F. Singer, Galois Theory of Difference Equations. VII, 179 pages. 1997.

Vol. 1667: J. M. F. Castillo, M. González, Three-space Problems in Banach Space Theory. XII, 267 pages. 1997.

Vol. 1668: D. B. Dix, Large-Time Behavior of Solutions of Linear Dispersive Equations. XIV, 203 pages. 1997.

Vol. 1669: U. Kaiser, Link Theory in Manifolds. XIV, 167 pages. 1997.

Vol. 1670: J. W. Neuberger, Sobolev Gradients and Differential Equations. VIII, 150 pages. 1997.

Vol. 1671: S. Bouc, Green Functors and G-sets. VII, 342 pages. 1997.

Vol. 1672: S. Mandal, Projective Modules and Complete Intersections. VIII, 114 pages. 1997.

Vol. 1673: F. D. Grosshans, Algebraic Homogeneous Spaces and Invariant Theory. VI, 148 pages. 1997.

Vol. 1674: G. Klaas, C. R. Leedham-Green, W. Plesken, Linear Pro-p-Groups of Finite Width. VIII. 115 pages. 1997.

Vol. 1675: J. E. Yukich, Probability Theory of Classical Euclidean Optimization Problems. X, 152 pages. 1998.

Vol. 1676: P. Cembranos, J. Mendoza, Banach Spaces of Vector-Valued Functions. VIII, 118 pages. 1997.

Vol. 1677: N. Proskurin, Cubic Metaplectic Forms and Theta Functions. VIII, 196 pages. 1998.

Vol. 1678: O. Krupková, The Geometry of Ordinary Variational Equations. X, 251 pages. 1997.

Vol. 1679: K.-G. Grosse-Erdmann, The Blocking Technique. Weighted Mean Operators and Hardy's Inequality. IX, 114 pages. 1998.

Vol. 1680: K.-Z. Li, F. Oort, Moduli of Supersingular Abelian Varieties. V, 116 pages. 1998.

Vol. 1681: G. J. Wirsching, The Dynamical System Generated by the 3n+1 Function. VII, 158 pages. 1998.

Vol. 1682: H.-D. Alber, Materials with Memory. X, 166 pages. 1998.

Vol. 1683: A. Pomp, The Boundary-Domain Integral Method for Elliptic Systems. XVI, 163 pages. 1998.

Vol. 1684: C. A. Berenstein, P. F. Ebenfelt, S. G. Gindikin, S. Helgason, A. E. Tumanov, Integral Geometry, Radon Transforms and Complex Analysis. Firenze, 1996. Editors: E. Casadio Tarabusi, M. A. Picardello, G. Zampieri. VII, 160 pages. 1998.

Vol. 1685: S. König, A. Zimmermann, Derived Equivalences for Group Rings. X, 146 pages. 1998.

Vol. 1686: J. Azéma, M. Émery, M. Ledoux, M. Yor (Eds.), Séminaire de Probabilités XXXII. VI, 440 pages. 1998.

Vol. 1687: F. Bornemann, Homogenization in Time of Singularly Perturbed Mechanical Systems. XII, 156 pages. 1998.

Vol. 1688: S. Assing, W. Schmidt, Continuous Strong Markov Processes in Dimension One. XII, 137 page. 1998.

Vol. 1689: W. Fulton, P. Pragacz, Schubert Varieties and Degeneracy Loci. XI, 148 pages. 1998.

Vol. 1690: M. T. Barlow, D. Nualart, Lectures on Probability Theory and Statistics. Editor: P. Bernard. VIII, 237 pages. 1998.

Vol. 1691: R. Bezrukavnikov, M. Finkelberg, V. Schechtman, Factorizable Sheaves and Quantum Groups. X, 282 pages. 1998.

Vol. 1692: T. M. W. Eyre, Quantum Stochastic Calculus and Representations of Lie Superalgebras. IX, 138 pages. 1998.